T0320177

The Digitalisation of Anti-Corruption in Brazil

This book investigates how digital technologies, such as social media and artificial intelligence, can contribute to combating corruption in Brazil.

Brazil, with its long history of scandals and abundant empirical data on digital media usage, serves as a perfect case study to trace the development of bottom-up and top-down digital anti-corruption technologies and their main features. This book highlights the connections between anti-corruption reforms and the rapid implementation of innovative solutions, primarily developed by tech-savvy public officials and citizens committed to anti-corruption efforts. The book draws on interviews with experts, activists, and civil servants, as well as open-source materials and social media data, to identify key actors, their practices, and the challenges and limitations of anti-corruption technologies. The result is a thorough analysis of the digitalisation of anti-corruption in Brazil, with a theoretical framework that can also be applied to other countries. The book introduces the concept of "integrity techies" to encompass social and political actors who develop and facilitate anti-corruption technologies, and discusses different outcomes and issues associated with digital innovation in anti-corruption.

This book will be a key resource for students, researchers, and practitioners interested in technologies and development in Brazil and Latin America, as well as corruption and anti-corruption studies more broadly.

Fernanda Odilla holds a PhD in Social Science and Public Policy and a MA in Criminology from King's College London. She is currently a lecturer in the Department of Political and Social Sciences at the University of Bologna and a work package leader for the RESPOND (Rescuing Democracy from Political Corruption in Digital Societies) project, funded by the European Union's Horizon Research and Innovation Action (RIA) programme. Odilla is also an associate researcher for the BIT-ACT (Bottom-Up Initiatives and Anti-Corruption Technologies) project supported by the European Research Council. Before her academic career, she worked as a multimedia producer for the Brazilian desk at the BBC World Service in London and as a reporter for daily newspapers in Brazil, where she was dedicated to investigating and exposing corruption.

Routledge Corruption and Anti-Corruption Studies

The series features innovative and original research on the subject of corruption from scholars around the world. As well as documenting and analysing corruption, the series aims to discuss anti-corruption initiatives and endeavours, in an attempt to demonstrate ways forward for countries and institutions where the problem is widespread. The series particularly promotes comparative and interdisciplinary research targeted at a global readership.

In terms of theory and method, rather than basing itself on any one orthodoxy, the series draws broadly on the tool kit of the social sciences in general, emphasizing comparison, the analysis of the structure and processes, and the application of qualitative and quantitative methods.

Corruption and Development in Nigeria
Edited by Ọláyínká Àkànle and David O. Nkpe

Corruption Proofing in Africa
A Systems Thinking Approach
Edited by Dan Kuwali

Corruption, Ethics, and Governance in South Africa
Issues, Cases, and Interventions
Edited by Modimowabarwa Kanyane

Deconstructing Corruption in Africa
Edited by Ina Kubbe, Emmanuel Saffa Abbdulai and Michael Johnston

The Digitalisation of Anti-Corruption in Brazil
Scandals, Reforms, and Innovation
Fernanda Odilla

For more information about this series, please visit: www.routledge.com/ Routledge-Corruption-and-Anti-Corruption-Studies/book-series/RCACS

The Digitalisation of Anti-Corruption in Brazil

Scandals, Reforms, and Innovation

Fernanda Odilla

Routledge
Taylor & Francis Group

LONDON AND NEW YORK

First published 2025
by Routledge
4 Park Square, Milton Park, Abingdon, Oxon OX14 4RN

and by Routledge
605 Third Avenue, New York, NY 10158

Routledge is an imprint of the Taylor & Francis Group, an informa business

British Library Cataloguing-in-Publication Data
A catalogue record for this book is available from the British Library

ISBN: 978-1-032-35380-7 (hbk)
ISBN: 978-1-032-35384-5 (pbk)
ISBN: 978-1-003-32661-8 (ebk)

DOI: 10.4324/9781003326618

Typeset in Times New Roman
by Taylor & Francis Books

Contents

Illustrations

Figures

Tables

Acknowledgements

This monograph is part of the BIT-ACT (Bottom-Up Initiatives and Anti-Corruption Technologies) project conducted by the Department of Political and Social Sciences at the University of Bologna, Italy. The project received funding from the European Research Council under the European Union's Horizon 2020 research and innovation programme (grant agreement no. 802362). The project's website is available at https://site.unibo.it/bit-act/en.

I express my gratitude to the BIT-ACT team for their insights, which helped me to put things into perspective. Special thanks go to the BIT-ACT's PI, Professor Alice Mattoni, for giving me the opportunity to delve into the field of anti-corruption technologies and for always providing valuable feedback. I would also like to thank Clarissa dos Santos Veloso, whose assistance in collecting data in Brazil and sharing the workload of transcribing interviews was greatly appreciated. Thanks to the comments and questions of Professors Lucio Picci, Nils Köbis, Armando Castro, and Luciano da Ros, who kindly read this monograph at such short notice, I was able to improve my analysis. I would also like to thank João Maciel and Fernando Castro, my tech-savvy friends, for helping me to craft a more accurate and easily readable academic monograph.

1 Introduction

When Aldous Huxley merged science with literature in his 1932 dystopian novel *Brave New World*, it was difficult to predict the impact his work would have. Although the novel was successful in terms of sales, reviews were consistently negative, with Huxley being accused of being "dry and boring," and his vision of the future was considered irrelevant and unoriginal (Bloom, 2004, p. 12). Over time, however, not only did Huxley's cautionary portrayal of the future become more connected to reality, but his critique of technology as a cure-all for issues stemming from wars and diseases found increasing resonance with his readers. Whether used ironically or not, the phrase "brave new world" is synonymous with a new context, often marked by socio-technological shifts and characterised by uncertainty regarding its potential success or benefit.

One could argue that evoking the phrase "brave new world" when discussing the digitalisation of anti-corruption efforts related to law enforcement and civic action, as is the case in this book, is somewhat clichéd. However, to a significant degree, the development and use of technologies in anti-corruption, especially emerging technologies, can be understood through Huxley's critical approach. This is because anti-corruption technologies represent a scenario where prevailing optimism often outweighs critical views and the necessary caution.

That is why the pages that follow can be seen as an attempt to critically investigate how a wide range of digital technologies, from social media platforms to artificial intelligence (AI) and blockchain applications, has been developed and employed to counter corruption by exploring the case of Brazil. This is a country with a high level of corruption, incremental accountability mechanisms (Da Ros and Taylor, 2022), and the largest number of democratic innovations in Latin America, in particular those designed by governments and civil society alike to enhance citizens' participation in anti-corruption efforts

DOI: 10.4324/9781003326618-1

and ensure good governance, often through the use of digital technologies (Pogrebinschi, 2018).

This is not to say that the book takes current and future technological developments to their extremes, juxtaposing inventions between insane utopia and barbaric lunacy, as Huxley did when depicting the choices of his characters. There is no need to create a dystopia nor to deny technological advances to maintain a shrewd outlook on technology in the context of anti-corruption efforts. What is necessary, then, is an analysis grounded in empirical evidence to better assess anti-corruption technologies, along with the contexts facilitating their development as well as their outcomes, limits, and risks. Assessments of technologies designed to counter corruption are emerging fields that still lack substantial research.

What do we know so far about anti-corruption technologies?

It is widely acknowledged that technological advances have led to unprecedented access to many different types of digital (and digitalised) data and innovative digital media that can rapidly be produced and used by various types of actors, located in both the public and private sectors, including citizens' associations and civil society organisations (CSOs). The surge in accessibility to high-volume and diverse micro-level data, combined with the widespread adoption of digital media and new data processing and analytical tools, has heightened expectations concerning the detection, prevention, and countering of corruption, an issue that international organisations have highlighted as a paramount global concern since the early 1990s.

However, it was only in the late 1990s and early 2000s that technologies were specifically used to fight corruption, mostly in the form of e-government projects (Kossow, 2020b). South Korea holds a prominent position in this context, with the Seoul Metropolitan Government implementing e-government reforms from the mid-1990s onwards; later, they were recognised as an effective tool against corruption (Iqbal and Seo, 2008; OECD, 2016). In subsequent years, organisations and researchers have delved into the use of technology in combating corruption, examining the topic broadly (Sturges, 2004; Bertot et al., 2010; Davies and Fumega, 2014; Kukutschka, 2016; Kossow and Kukutschka, 2017; Mattoni, 2017; Adam and Fazekas, 2018, 2021; Kossow, 2020a), or focusing on specific types of technologies such as AI (Aarvik, 2019; Köbis et al., 2022a; Odilla, 2023a), social media (UNDP, 2011; Jha and Sarangi, 2017), crowdsourcing (Noveck et al., 2018; Zinnbauer, 2015), e-government (Andersen, 2009;

Elbahnasawy, 2014), open data (Gurin, 2014), distributed ledger technology like blockchain (Kim and Kang, 2019; Kossow and Dykes, 2018), or specific devices such as mobile phones (Chêne, 2012). In addition, there were fewer attempts to focus on the issues revolving around emerging technologies, such as the corruption risks of AI (Köbis et al., 2022b) or the risk of unfairness when creating AI-based anti-corruption tools (Odilla, 2023b).

Overall, the role of digital technologies in fighting corruption has become a topic that is usually covered by international organisations, such as the U4 Anti-Corruption Research Centre, Transparency International, the Organisation for Economic Co-operation and Development, and the World Bank, to name but just a few. In addition, the academic literature on anti-corruption digital technologies (ACTs) is generally scant; reports and policy papers are more common. It is rarely based on exhaustive empirical research and instead relies heavily on anecdotal evidence. Indeed, in the academic field, publications are mostly based on secondary data (see Köbis et al., 2022a; Mattoni, 2021; Adam and Fazekas, 2021).

There are noteworthy exceptions in terms of empirical data, such as the mapping of Brazilian AI-based tools showcasing data inputs, data processing and outputs, as well as the prevalent types of corruption that these technologies aim to address (Odilla, 2023a), and diagnosis of the unfairness of risk estimation tools for public contracts used by law enforcement agencies in Brazil and the implementation of mitigation measures (Lima and Andrade, 2019). This is also the case in a survey undertaken in Germany by Starke et al. (2023) which scrutinised the best design of an automated Twitter bot to foster collective action. The book *Digital Media and Grassroots Anti-Corruption*, edited by Alice Mattoni (2024), can also be seen as a significant exception in that it brings together empirical cases presented by different authors from around the world. However, the book focuses only on bottom-up anti-corruption technologies, without assessing government tools. Meanwhile, governments have embraced a range of digital tools and applications primarily aimed at verifying public procurement and bidding processes, flagging anomalies in social benefits payments, and bolstering trust while automating financial records and transactions to curb corruption (Odilla, 2023a, 2021).

When discussing anti-corruption technologies in Brazil, most analyses consist of isolated case studies, with the majority focusing on civil society initiatives or specific types of technologies. For example, Odilla (2023a) drew up a list of over 30 bottom-up and top-down cases but concentrated only on those utilising AI. Still, one important finding is the very low level of concern among developers of anti-corruption tools about risks such as bias and unfairness, as well as about having auditable systems in the anti-corruption realm. Neves et al. (2019), in turn, evaluated how civil servants from one specific government agency, the Federal Court of Accounts, are approaching the digitalisation of anti-corruption law enforcement. In another study, the focus was on the use of social media to promote two anti-corruption grassroots campaigns aimed at passing new legislation to combat corruption, suggesting that a key feature of successful change in public policies is the result of collectively organised agents deploying innovative strategies to exert public pressure, which can sometimes surprise politicians in their electoral cost-benefit analyses (Mattoni and Odilla, 2021).

Freire et al. (2020) and Galdino et al. (2023), in turn, have investigated bottom-up accountability strategies through interventions designed to enhance ordinary citizens' monitoring capacity regarding public services. The former presented experimental evidence that a mobile app (*Tá de Pé*) provided to citizens did not improve the delivery of school and nursery construction works, while the latter evaluated a related project called *Obra Transparente*, developed by the non-governmental organisation (NGO) *Transparência Brasil,* formerly part of Transparency International and operating independently in the country since 2007. They concluded that costly mobilisation and monitoring efforts by organised CSOs are more likely to drive significant policy change than less costly projects such as the development of apps that can be used by anyone with a mobile phone connected to the internet. This finding suggests that success is less about the technology itself and more about the mobilising efforts required to bring people together to effectively utilise anti-corruption technology.

Even with the emergence of multiple top-down and bottom-up initiatives aimed at curbing corruption through the use and development of digital technologies, not only in Brazil but also around the globe, we still lack a comprehensive knowledge of how anti-corruption technologies emerge, their main practices and how they bolster the fight against corruption. By focusing on major technological advances experienced by Brazil at the federal level, this book explores the digitalisation of anti-corruption, showing that the country has been rapidly incorporating different types of digital technologies, most of them developed in-house by tech-savvy civil servants and concerned citizens, to deploy in the government's anti-corruption initiatives. This is to be expected, not only because digital technologies play a big role in our lives but also because they can lower costs and barriers to people's engagement and participation in anti-corruption efforts (Earl and

Kimport, 2011; Bennett and Segerberg, 2013) and speed up processes to support human activities given the expectation that machines are immune to fatigue (Köbis et al., 2022a).

However, it is safe to say that it is not sufficient for digital technologies merely to exist within societies for use in anti-corruption efforts. There is much yet to be explored, with important questions needing to be addressed. What circumstances can lead to the emergence of ACTs? Which of the most prevalent elements of ACTs are already in place? How exactly do they function and how can we measure their outcomes? Who are the primary human and non-human actors developing and facilitating the use of digital technologies as tools against corruption? What are their main practices, and what are their principal challenges and limitations? Can ACTs provide an antidote to corruption, or are they more likely to exacerbate discredit, thus fuelling democratic regression rather than creating awareness and prompting indignation (and action)? In seeking to tackle these and other fundamental questions in this burgeoning field of inquiry, this volume examines the significant roles that digital technologies may play in combating corruption, paying heed to both human and non-human actors.

Decoding key concepts

Before proceeding further, it is crucial to establish the conceptual groundwork of this book. While new concepts and typologies will be introduced throughout this book, there are key definitions that need to be clarified in advance, starting with corruption. Although corruption is a contested concept, there is growing consensus that it is multifaceted, varying in types and intensity, evolving over time, and it is not exclusive to immature democracies or the Global South. Hence, corruption is understood here as an umbrella concept encompassing various types of conduct related to the misuse of power for private gain at the expense of the collective, incorporating concepts such as clientelism, patronage, patrimonialism, state capture, and particularism (Varraich, 2014). As Michael Johnston (2005, p. 11) defines it, "corruption involves the abuse of a trust, generally one involving public power, for private benefit which often, but by no means always, comes in the form of money."

Despite being a complex problem that still challenges those who aim to understand and curb it, this book assumes that corruption cannot be solved solely through the use of technical solutions. The scant existing research on digital technologies suggests that, despite their often innovative features and, therefore, the considerable media buzz they initially attract, most of these technologies disappear shortly after their launch (Kukutschka, 2016). Digital technologies are treated here as a diverse array of technological resources, including data, data-processing techniques, software and hardware, deployed to electronically create, distribute, view, and store digital information that governments, activists, and concerned citizens might use to sustain their actions and their communications repertoire (Mattoni, 2017). Digital technologies are recognised as valuable resources in anti-corruption efforts, but it must be acknowledged that they require specific means and capacities, including financial resources, tech literacy and openness to digital innovation, if they are to be developed and employed to support anti-corruption efforts (Odilla, 2024). This view may help to explain why many anti-corruption activists, for example, still rely on analogic actions (Odilla, 2024) and why many anti-corruption technologies are reluctantly embraced by frontline workers in public administration in Brazil (Neves et al., 2019).

ACTs are defined as complex systems designed with the overarching aim of combating corruption, as described by Mattoni (2024). ACTs are assemblages of human and non-human actors who simultaneously pursue immediate and practical anti-corruption objectives, addressing directly or indirectly various levels of corrupt activities ranging from petty to grand corruption-related wrongdoings. The non-human ACTs' components are digital data, algorithms, and hardware, and they do not exist in a void. ACTs also have a human component, including developers, facilitators, and users, each with their own imaginaries on corruption, anti-corruption efforts, and the deployment of technology in combating corruption. While digital technologies are part of the ACTs' *material* dimension, humans and their social relationships are seen as *social*, and the *symbolic* elements are the imaginaries that sustain the creation and use of ACTs in anti-corruption practices (Mattoni, 2024). ACTs, therefore, are not synonymous with digital technologies. They go beyond them.

Thus, this book combines Lascoumes and Le Galès' (2007) insights on the instrumentation of public policy and the definition of ACTs presented by Mattoni (2024) when reflecting on civil society initiatives to assess the top-down and bottom-up ACTs already in place in Brazil. In line with these authors, the assumption here is that ACTs cannot be considered solely as functional instruments or pragmatic solutions for achieving more efficient results because they are neither denaturalised technical objects nor neutral devices. It means that they cannot be considered solely from an instrumental viewpoint because ACTs are a combination of material, social, and symbolic elements, as proposed by Mattoni (2024).

As with corruption, ACTs are embedded in politics and power asymmetries. The aim here is to look at the digital technologies deployed in anti-corruption activities, while at the same time making an effort not to see technologies as isolated or disconnected from humans, or from policymaking and law enforcement. This book, therefore, adds a political dimension to the concept of ACTs, alongside the material, social, and symbolic elements already mentioned. The political dimension encompasses the political context and legal apparatus, as well as their influence on the duties and the agency of those responsible for developing or facilitating the development of digital technologies. These aspects, as well as the importance of agency in socio-technical change, are as important as the technological interventions themselves.

Humans and non-humans in the "web of accountability"

To guide the analysis of recent technological changes in the anti-corruption domain, this book revisits the "web of accountability" concept (Mainwaring and Welna, 2003). The web of accountability, as defined by Mainwaring and Welna (2003) and Power and Taylor (2011), refers to a network of institutions comprising the mechanisms of accountability, encompassing the interplay among these institutions, as well as the interaction between electoral accountability, intra-state accountability, and societal oversight (Mainwaring and Welna, 2003). Although initially introduced to examine democratic accountability in Latin America and already deployed to explain the accountability system in Brazil (Carson and Mota Prado, 2014; Power and Taylor, 2011; Aranha, 2020), as an analytical approach it is widely applicable, describing the interconnected system through which various actors assume oversight, investigative, and disciplinary roles. Overall, anti-corruption systems include a broad range of governmental and non-governmental actors with complementary and compensatory roles aimed at holding governments, public spending and service delivery, and public officials accountable. I expand this understanding, and consider both human and non-human actors as part of this web.

Accountability is understood here as the answerability and responsibility of public officials (Mainwaring and Welna, 2003). Following O'Donnell's (1999) definition, it assumes horizontal and vertical relationships. Horizontal accountability is seen as

> the existence of state agencies that are legally enabled and empowered, and factually willing and able, to take actions that span from routine overseeing to criminal sanctions or impeachment regarding

actions or omissions by other agents or agencies of the state that may, presumably, be qualified as unlawful.

(O'Donnell, 1998, p. 11)

While horizontal accountability includes both internal and inter-agency control in the legislature, executive, and judiciary branches, vertical accountability is viewed as an evolving overarching concept encompassing a variety of actions coming from non-state actors. It includes citizens voting based on elected officials' actions in office (electoral accountability); citizen-led monitoring and scrutiny of public and/or private sector performance, including the press and CSOs that investigate and denounce abuses and wrongdoings (societal accountability); and user-centric access and distribution systems for public information and citizen involvement in tangible decision-making related to resource allocation, such as participatory budgeting (social accountability) (Grimes, 2008; Smulovitz and Peruzzotti, 2000).

While acknowledging previous scholarly efforts to define various types of accountabilities, this book adopts a more simplified approach. Thus, accountability is divided into two clusters: top-down and bottom-up. The former refers to the accountability of government actors, and the latter to the accountability of non-government actors (including, but not limited to, individual concerned citizens, CSOs, and journalists), respectively. As digitalisation and automation processes advance, non-human entities also play increasingly significant roles in various phases of the accountability cycle. As emphasised by Mota Prado et al. (2015) and Odilla and Rodriguez-Olivari (2021), the accountability cycle employs three primary functions:

1 *Oversight* involves the active monitoring of activities with a significant risk of corruption in order to promptly prevent and/or identify any suspicious or unusual occurrences.
2 *Investigation* is the systematic process of gathering comprehensive information about specific actions or activities once suspicions have been raised.
3 *Corrective measures* ensure the enforcement of sanctions in cases where there is conclusive evidence to substantiate misconduct.

While this typology was provided mainly to assess government institutions responsible for horizontal accountability (i.e. control within and among government agencies, as described by O'Donnell, 1998), these functions are equally applicable to the process of bottom-up accountability. In addition, they encompass situations where civil servants, civil society actors, and their digital technologies provide support to each other's actions or act independently. Here, ACTs are not seen solely as enablers of accountability practices but as part of the web of accountability. We must visualise ACTs as nodes within the web of accountability, where human and non-human actors have overlapping, competing, or complementary roles. ACTs are, in turn, assemblages – sometimes temporary, sometimes more long-lasting – of social, symbolic, material, and political elements. The responsibilities and goals of both digital technologies and their developers and facilitators may vary depending not only on the primary function of their actions but also on how technologies are imagined, designed, developed, and used. This implies that various law enforcement agents and CSOs, for example, have distinct priorities and dedicated responsibilities, which may lead them to concentrate on specific or broader anti-corruption actions. Similarly, the digital technologies they deploy may serve single or multiple purposes. This rationale guided both the gathering and analysis of the data collected in this study.

Research design, methods, and data analysis

The research for this book was conducted using the framework developed by the BIT-ACT (Bottom-Up Initiatives and Anti-corruption Technologies) project.[1] Methodologically, the monograph follows the project's qualitative approach, which combines constructivist grounded theory with situational analysis (Charmaz, 2006; Clarke, Friese and Washburn, 2015; Mattoni, 2020). This methodological approach places value on the perspective of activists, concerned citizens, and civil servants who have been engaging in anti-corruption efforts, mainly by developing and supporting the creation and use of digital technologies to combat corruption. The focus is on ACTs, which have served as a sensitising concept (Bowen, 2019) for the empirical research in this book, as well as for the aforementioned BIT-ACT project. This is why the analysis is guided through the lens of ACTs, as both this book and its empirical research are firmly rooted in the research project.

Brazil is seen here as the context where the creation, development, and use of ACTs occur. Context matters and, therefore, the research process has employed several steps to capture its nuances. The first step was to conduct desk research and expert interviews to understand how corruption manifests itself, who the main actors fighting corruption are, and which ACTs are already in place in Brazil; indeed, it is worth stressing that it was the search for anti-corruption initiatives in which ACTs had a relevant role that guided the selection of case studies and the subsequent fieldwork that was undertaken.

The second step involved the creation of situational maps based on specific initiatives and their digital technologies. These situational maps depicted key human, non-human, material, symbolic, social, and discursive elements in the situations of concern (Clarke et al., 2015). The situational analysis approach was especially valuable for gaining initial insights into how different actors were interacting, and which types of technologies have been deployed. Situational maps were also employed to choose the initial set of case studies and to position them within the various arenas, each in its specific context, as well as to identify who should be invited to participate in this study.

The third step involved gathering data on the first group of selected case studies by carrying out in-depth semi-structured interviews, participant online and offline observations, by collecting other materials such as visuals, reports, news media articles, and by social media scraping. These data underwent an initial round of qualitative analysis following the core principles of grounded theory, which involved a combination of open and focused coding (Charmaz, 2006). The analytical process was complemented by the theoretical sampling of the initial data, whereby subsequent rounds of data gathering and analysis were conducted based on the findings that emerged from the data already collected (Glaser and Strauss, 1967). These rounds involved the collection and analysis of additional data. The MAXQDA Plus 2020 software package was used for the analysis, as well as all the different types of coding.

This interactive fieldwork approach proved invaluable for selecting and then excluding a few initially selected cases, including new ones, exploring emerging research topics, and enhancing the data collection. The case studies selected for investigation are outlined in Chapters 3 and 4 and summarised in the online Appendix.[2] They offer significant variations that are valuable for obtaining a more comprehensive view of how anti-corruption efforts evolve and/or adapt digital technologies from both the bottom-up and the top-down, going beyond simply anecdotal evidence. The case studies vary in terms of the type and ownership of the digital technologies employed, ranging from existing social media platforms to emerging technologies such as AI and blockchain developed for specific purposes. They also differ in terms of the type of corruption targeted, the point when these technologies began to be utilised, and the types of actors involved in the anti-corruption initiatives.

The overall aim here is not to map all the existing initiatives in Brazil but to present some representative case studies in order to conduct an in-depth analysis of the multifaceted employment of digital technologies in anti-corruption efforts. The resulting sample contains, in some cases, specific ACTs as the unit of analysis, such as the X (Twitter) bots Rui from the media outlet JOTA and Rosie from *Operação Serenata de Amor* (Operation Love Serenade). In other cases, not only are the ACTs considered but also CSOs, such as *Transparência Brasil* and *Operação Política Supervisionada*, as well as law enforcement agencies, such as the *Corregedoria Geral da União* (Office of the Comptroller General), the *Tribunal de Contas da União* (Federal Court of Accounts), and the *Receita Federal* (Revenue Service) due to the significant number of digital technologies they have been developing. Some cases started offline and later incorporated digital tools in order to pursue their goals, some cases were born digital and presented a co-evolutionary path interplaying with technology developments, some cases simply make use of social media or were embedded in the most popular technology at the point when they were created (e.g. chatbots on specific social media platforms, transparency portals, desktop and mobile oversight applications), or even more innovative experiences that deploy blockchain and AI-based ACTs.[3]

The case studies selected follow suggestions by Odilla (2023) and Köbis et al. (2022) that there should be separate approaches for civil society and government-led anti-corruption technologies. For analytical purposes, we do not consider the size or cost of the initiatives as criteria for clustering the case studies. The focus is exclusively on the actors responsible for developing and/or adapting digital technologies to be deployed in anti-corruption efforts. Therefore, the selected case studies can be divided into two main categories, as shown below.

Bottom-up anti-corruption efforts

These are initiatives led by CSOs, collective actors, or concerned citizens, who develop their own digital technologies or utilise existing platforms (like social media) to advance their anti-corruption agendas.

Among the bottom-up initiatives there are two main types of case studies. First, there are movements or collective actions that have employed digital technologies to raise awareness and mobilise people to counter corruption through online and offline demonstrations and broad campaigns that seek to produce change at the executive and legislative level and, hence, obtain policy outcomes. These initiatives follow a logic of collective actions to pursue their goals and integrate digital media to increase citizen participation and campaign visibility. Second, there are informal groups of citizens and NGOs gathered around specific anti-corruption initiatives that use digital technologies intended to augment the monitoring capabilities of people, primarily through the use of open public information about corrupt practices or related misconduct. These initiatives follow a logic of connective actions (Bennett and Segerberg, 2013) which put individuals connected through digital media at the centre of anti-corruption initiatives.

Top-down anti-corruption efforts

These are initiatives led by state actors, primarily civil servants working for anti-corruption and other law enforcement agencies, that involve the development and deployment of digital technologies to automate and accelerate procedures that can help humans to make decisions and conduct their daily tasks or that can offer citizens channels for accessing public information, receiving information on corruption, or reporting instances of corruption.

There are three main types of case studies among the top-down anti-corruption efforts. First, there are systems, most of them automated, that are used to monitor, identify, or predict suspicious cases. The type of corruption they target is directly related to the duties of the agencies where these digital technologies were developed or are being deployed. Second, government initiatives interact with ordinary citizens to crowdsource information and/or offer anti-corruption-related services. Third, some top-down initiatives are undertaken that favour bottom-up actions. State actors, many of them law enforcement authorities, engage in grassroots anti-corruption efforts in different capacities, such as volunteering training on how to use open data, organising events such as hackathons or workshops with civil society actors, and leading campaigns to improve the legal anti-corruption apparatus.

Top-down initiatives do not necessarily involve large and expensive projects, as will be shown. They benefit from previous more cost-conscious governmental initiatives that have digitised public data and made them accessible in machine-readable formats, with most of them – the open public data – also serving as raw material for bottom-up initiatives. As will be explored, most bottom-up and top-down initiatives do not outsource their solutions. In Brazil, digital anti-corruption initiatives are mainly an "in-house" innovation process, undertaken by tech-savvy and non-tech-savvy individuals alike, all of whom are more open to digital transformation.

The participants, therefore, were selected based on their roles in each type of initiative. There were six different types of interviewees:

1 *Grassroots initiators*: tech-savvy and non-tech-savvy individuals, among them activists, concerned citizens, and CSO representatives, directly participating in the creation and/or deployment of ACTs.
2 *State actors as initiators/facilitators*: public officials directly involved in anti-corruption campaigns and/or in the creation and use of ACTs.
3 *IT designers and developers*: individuals hired to build tech solutions.
4 *Collaborators*: people who assist in the development and/or strategic use of ACTs by serving as volunteer tech developers or supporting initiatives financially or with ideas.
5 *Journalists* using, creating, or promoting ACTs.
6 *Followers of ACTs* on social media who may or may not have used the technologies.

As expected, not all case studies involved individuals playing all these roles. The number of interviews, therefore, varied according to the different types of participants in each case study as well as the saturation point. In total, 77 participants, including five experts, were interviewed between March 2020 and April 2023, either online or offline. Online interviews were carried out using each interviewee's platform of preference, mainly Zoom and Google Meet, and face-to-face interviews were conducted in Brazil (São Paulo and Brasília) and Switzerland. All the interviews were conducted in Portuguese: most of them were recorded in audio and others were conducted in written format. The participants' names were converted into alphanumeric numbers for anonymisation. Table 1.1 shows the gender, age cohort, and type of participants, along with the frequency and percentage of their demographics.

The interviewees were asked to discuss, among other topics, their motivations and goals in engaging in anti-corruption efforts, their views on using data and technology to combat corruption, and their overall assessment of the tools they had created and/or used, including the main challenges, outcomes, and their views on what users of ACTs want. The interviewees were also asked to define what corruption is. Both the in-depth interviews and the other data gathered were analysed abductively. The result was an in-depth analysis of the fertile ground for digital innovation in the anti-corruption realm in Brazil, the development of the concept of "integrity techies" to define the wide range of social and political actors who harness digital technologies to strengthen public integrity, as well as new typologies related to outcomes, the main features, and the risks associated with digital technologies in anti-corruption. Although both the data and analyses

Table 1.1 Demographics of the interviewees

	Frequency	*Percentage*
Gender		
Male	55	71.4
Female	22	28.6
Age cohort		
20–29 years	7	9.1
30–39 years	30	38.9
40–49 years	16	20.8
50–59 years	5	6.5
60–69 years	3	3.9
70–79 years	1	1.3
80–89 years	2	2.6
Missing values	13	16.9
Participants		
Grassroots initiators	27	35.1
Hired IT designers/developers	14	18.2
ACT followers	7	9.1
State actors as initiators/facilitators	21	27.3
Journalists	3	3.9
Experts (in anti-corruption, social movements and civil society, and ICT/digital technology)	5	6.4

in this book tell the story of Brazil, what emerged from the vast quantity of empirical evidence under study not only gives us a better understanding of the circumstances under which ACTs are more likely to arise but also facilitates a closer examination of the limitations and risks and the implication thereof to the outcome of anti-corruption efforts.

Outline of the book

Over the past decade, an increasing number of studies have focused on examining the social, material, political, and organisational dimensions of the technological infrastructures that underpin global information and communication flows (Bonini and Trerè, 2024, p. 7). However, with very few exceptions (e.g. Mattoni, 2024; Odilla, 2023a; Starke et al., 2023; Odilla and Mattoni, 2023; Köbis et al., 2022a; Ceva and Jiménez, 2022;

Adam and Fazekas, 2021; Mattoni and Odilla, 2021; Kossow, 2020a, 2020b), there has been limited reflection on the application of these aspects to anti-corruption efforts. More importantly, as noted above, there has been little empirical research on the application of these dimensions, taking into account the context, the human and non-human actors involved, and their outcomes, obstacles and risks.

This book builds on robust empirical evidence to assess the key elements of ACTs individually, and to provide an analysis of these socio-technical-political assemblages implemented in Brazil. It first presents the political and legal context related to the gradual strengthening of the accountability network and digital transformation in Brazil. It then assesses the digital technologies developed and used in anti-corruption initiatives, identifying the people involved in these efforts and their main social and political roles considering the accountability network. Next, the book adopts the practice theory approach (Mattoni, 2017) to analyse the ACTs, where material, symbolic, social, and political elements mutually shape each other. Finally, the book explores the limitations and challenges of ACTs and how this affects their outcomes.

Accordingly, Chapter 2 provides a context that is not only pertinent for analysing the case of Brazil but also for laying the groundwork for future comparisons. It begins by outlining the various shapes and forms that corruption takes in the country, tracing Brazil's decades-long history of corruption scandals, as well as the promises of anti-corruption embedded in the political discourse. It also presents grassroots and government efforts that have led to the incremental strengthening of the web of accountability institutions and the digitisation of anti-corruption processes in the country since the restoration of democracy in the mid-1980s. The detailed narrative of Brazil's digital transformation in public administration, encompassing the management of procedures, documents, and services, as well as its role in combating corruption, establishes a link between anti-corruption efforts and digital technologies. This connection is essential for identifying and explaining the significance of legal frameworks and the accountability apparatus in terms of policies and actions for digital innovation in the anti-corruption realm.

Chapter 3 provides an overview of the diverse types of digital technologies and discusses how and by whom they have been developed and used to combat corruption in Brazil. The chapter introduces what is termed "integrity techies": social and political actors who leverage digital technologies to bolster public integrity through anti-corruption, transparency, and accountability measures. The chapter presents the stories of several integrity techies and their digital technologies designed to curb corruption from both top-down and bottom-up perspectives. It highlights how the accessibility of open data, along with the level of digitalisation and transparency within the public sector in Brazil, has facilitated the development and implementation of digital technologies by civil servants, concerned citizens and CSOs, with very low levels of outsourcing.

Chapter 4 offers a more analytical perspective of the ACTs already in place in Brazil. Aiming to elucidate the proliferation of integrity techies and their digital technologies in the country, it first delineates the principal disparities between top-down and bottom-up initiatives from a situated and pragmatic standpoint. Subsequently, it delves into the material, symbolic, social, and political dimensions of ACTs. Within the sphere of anti-corruption efforts, there exists a reservoir of human resources poised to leverage their technological expertise to harness copious amounts of available data, craft new technologies, and use existing ones, a process facilitated by pertinent legislation and data availability. Furthermore, the analysis posits that financial constraints have compelled the adoption of in-house solutions. These aspects furnish a robust material groundwork for the advancement of ACTs in confronting what is perceived as pervasive systemic corruption. The chapter concludes by contemplating the key conditions under which integrity techies have emerged and proliferated in Brazil.

In Chapter 5 the focus is on the primary limitations and challenges associated with ACTs as well as with the digitalisation of anti-corruption efforts in Brazil. Despite the prevailing optimism surrounding the innovative landscape in the country, there are inherent risks that demand attention, particularly concerning the rapid evolution of technologies, including applications of AI, while lacking discussion on best practices and cutting-edge solutions for more effective and ethical anti-corruption efforts. The chapter raises key issues that vary from lack of clear metrics to evaluate outcomes, still limited access to standardised, machine-readable quality data, and difficulties in engaging users. Another critical area requiring attention is the high level of opacity of ACTs and unchecked technological advancement, heightening the risk of unfairness, bias, and noise in the fight against corruption.

Chapter 6 concludes the book by providing an upbeat overview of the remarkable and inventive efforts to bolster integrity, transparency, and accountability through digital technologies in Brazil. While acknowledging the constraints and hazards of ACTs in their present state there, this chapter ventures into the future, delving into the trajectory of digitalising anti-corruption efforts both within the country and globally. Insights gleaned from Brazil can play a pivotal role in fostering a more discerning outlook and establishing digital governance frameworks to facilitate the creation and endurance of more efficacious ACTs.

The pages that follow pave the way towards understanding ACTs, adding complexity to our perspective on the development and use of technologies in the context of anti-corruption.

It progresses from the political context to digital technologies, to a more intricate environment when considering ACTs and their four dimensions. This book is intended for researchers, policymakers, law enforcement agents, activists, tech-savvy individuals, and concerned citizens seeking actionable insights to implement technology-enabled anti-corruption solutions. It is imperative to involve all these actors to develop strategies to maintain vigilant oversight that will enable ACTs to effectively address the abuse of power for private gain at the expense of the collective good without perpetuating the problems we aim to solve. Furthermore, it is necessary to undertake a more critical examination of what both humans and digital technologies can do in the anti-corruption fight and to explore alternatives that are able to drive meaningful change.

I believe that this book will help to start filling those gaps, without being too dry or boring, as Huxley's work was once described. Jokes aside, I hope that all those interested in this "brave new world" will enjoy reading this book.

Notes

1 This research project was funded by the European Research Council under the European Union's Horizon 2020 research and innovation programme (grant agreement no. 802362).
2 F. Odilla (2024, March 19). *Anti-Corruption Technologies in Brazil: Online Appendix*. Available at https://osf.io/7gzwu/.
3 AI-based ACTs are defined here as "data processing systems driven by tasks or problems designed to, with a degree of autonomy, identify, predict, summarise, and/or communicate actions related to the misuse of position, information and/ or resources aimed at private gain at the expense of the collective good. This type of application implies the analysis of a given environment based on a set of predefined rules before acting" (Odilla, 2023a, p. 354).

2 "Inching" towards accountability and digital transformation

In 1954, before committing suicide in the wake of a corruption scandal, Brazilian President Getúlio Vargas was reported to have said: "I have the impression that I am on a sea of mud" (Long, 1988). Since then, the "sea of mud" metaphor has been used to describe Brazil's decades-long history of corruption scandals in all branches and levels of government (Power and Taylor, 2011; Lagunes et al., 2021a, 2021b; Da Ros and Taylor, 2022).[1] Over the years, the Brazilian political class has acquired a reputation of being corrupt, unruly, and enjoying a high degree of immunity (Mattoni and Odilla, 2021). Not by chance, Norberto Odebrecht, the founder of the Brazilian construction conglomerate Odebrecht,[2] once defined doing business with the government and interacting with politicians as "getting into the mud with the pigs but coming out clean and in a white suit" (Gaspar, 2020).

Ironically, according to the official public records of presidential inaugural speeches, Vargas was the first Brazilian president to explicitly mention the word "corruption" in an inaugural address (Lagunes et al., 2021b). In his 1930 speech, delivered after leading an armed revolt that placed him in power, Vargas listed many anti-corruption promises, including the promise to keep nepotism in check and to clean up the government.[3] After being ousted from office in 1945, Vargas returned to power as a democratically elected president in 1951. He took office with a renewed pledge to control corruption, only to find himself entangled in several corruption scandals that contributed to the demise of his government. Vargas's words call attention to how corruption is a longstanding issue in Brazilian politics. The same can be said about the often empty promises to fight corruption that became part of the political rhetoric in the country (Lagunes et al., 2021b).

In the case of Odebrecht, despite its founder's trope, the "white suit" eventually got covered in mud. Although the company's name had been mentioned in connection with various national scandals over the

DOI: 10.4324/9781003326618-2

years, such as the *Ferrovia Norte-Sul* (North-South Railroad) and *Anões do Orçamento* (Budget Dwarves),[4] it was only during the *Lava Jato* (Car Wash)[5] investigation that the company, its owners, and high-ranking directors were investigated and sanctioned for engaging in an unparalleled bid-rigging and bribery scheme (Gaspar, 2020). In 2016, Odebrecht and one of its petrochemical companies pleaded guilty to having paid approximately US $788 million in bribes to political parties from both sides of the spectrum, government officials and their representatives to win and keep business in 12 countries. They agreed to pay a fine totalling US $3.5 billion to resolve charges brought by authorities in Brazil, Switzerland, and the United States (DOJ, 2016).

The *Lava Jato* investigation, however, can be seen as a double-edged sword in the anti-corruption struggle. On the one hand, the operation launched in Brazil in 2014 became one of the world's largest criminal investigations into grand corruption due to its length, scope, the sums involved, the prominence of the defendants, the complex and long-lasting corruption schemes, and, finally, the ramifications thereof in the Americas and Africa (Da Ros and Taylor, 2022, p. 3). The investigation into the scandal outraged the country, fostered street protests in favour of *Lava Jato* and against the country's political elite (mainly members of the Partido dos Trabalhadores – Workers' Party), and provided fodder for multiple press headlines. The investigation also benefited from a previously unseen level of transparency. Both the advances in technology and decisions by judges and prosecutors meant that the public was able to access a plethora of raw material, including various case documents and videos of court hearings and plea-bargaining statements that were uploaded to the internet and made available within a few hours (Bulla and Newell, 2020). Law enforcers also deployed advanced technology, such as the Revenue Service's artificial intelligence (AI)-based tool known as *ContÁgil* that reads account books and invoices, scans the various data sources to which it has access, and builds network graphs to show the interconnections between people and companies (Odilla, 2023a).

On the other hand, in the aftermath of what at first glance seemed to be a very efficient level of coordination among law enforcement and accountability institutions in Brazil, numerous questions arose regarding the events that went on behind the scenes of *Lava Jato*. This was particularly true concerning the potential political bias of investigators and unethical exchanges between judges and prosecutors using the private messaging app Telegram, which became public in 2019 (Aranha, 2020, p. 185). This scandal became widely known as *Vaza Jato* (Jet Leak), with digital technology also playing a significant role in its unfolding. The messages were illegally obtained by a hacker who gained access to the Telegram accounts of the prosecutors involved in the case and Judge Sergio Moro and who then leaked a large amount of data to the news website Intercept Brasil, co-founded by Glenn Greenwald, who also helped to publish ex-CIA contractor's Edward Snowden's leaked documents on US surveillance programmes.

By 2021, seven years after *Lava Jato* was first brought to the attention of the public, not only had a number of landmark judicial decisions been overruled by the higher courts, but the investigation also lost popular support, its task force was dismantled, and the Brazilian anti-corruption apparatus was weakened (Lagunes et al., 2021b; Da Ros and Taylor, 2022). To make matters worse, *Lava Jato* did not seem to have reduced the level of political corruption nor to clean up politics as scandals involving politicians and businesspeople kept coming to light.

The importance of evaluating scandals and reforms

One way to better understand the rise and fall of the *Lava Jato* investigations in Brazil is by assessing the anti-corruption reforms adopted since democracy was restored in the country in 1985. Brazilian anti-corruption efforts, including the adoption of digital technologies as anti-corruption tools, have been implemented over the years mainly amid or as responses to specific corruption scandals. Hence, the accountability infrastructure was improved on demand and most often under domestic and international pressure. Revisiting some of these corruption scandals and how they made Brazil "inch toward accountability," as the slow and incremental institutional advances were described by Praça and Taylor (2014), also helps us to understand how the country developed an institutional apparatus that facilitated both the bottom-up and top-down use of digital media in the fight against corruption.

The anti-corruption advances and reforms made over the years were not, for example, able to establish robust mechanisms to guard the guardians themselves, curb assaults on the rule of law, or create safeguards to prevent backsliding in the anti-corruption battle. The multiple scandals and the key anti-corruption advances that resulted from them are essential to understanding the emergence of "integrity techies," i.e. social and political actors who have embraced digital technologies to fight corruption, increase transparency, and/or improve accountability in Brazil, and the bottom-up and top-down digital anti-corruption initiatives (both of which will be explored in detail in Chapter 3). In this context, the case study on Brazil corroborates findings from prior studies, with the literature suggesting that corruption scandals significantly influence institutional and civic responses (Bauhr and Grimes, 2014; Fernández-Vázquez et al., 2015; Wickberg, 2018).

The remainder of this chapter provides an overarching view of the manifestations of corruption in Brazil. It begins by exploring its shapes and forms, followed by a non-exhaustive list of scandals that have tarnished presidential administrations. This list serves as a brief history to contextualise the subsequent discussion on key institutional changes related to enhanced transparency, openness, accountability, and participation, and their connection to the digitalisation processes initiated since the promulgation of the 1988 Constitution in Brazil. Finally, it delves into the efforts to digitise anti-corruption efforts within the country.

Manifestations of corruption in Brazil: shapes and forms

In Brazil, as our brief glimpse of the "sea of mud" has already illustrated, scandals remain a troubling constant (Power and Taylor, 2011), particularly in its political system. A closer look at the most pre-eminent scandals also indicates a recurrent pattern of grand corruption (Rose-Ackerman, 1996) often associated with public procurement, political financing, tax administration, and natural resources, along with collusive behaviour between high-level figures and law enforcers. Although corruption is not exclusive to a model, form, or regime of government and has proved to be a multifaceted persistent phenomenon (Odilla, 2020b, p. 27), in Brazil corruption has been described as a systemic issue in which not only political and business elites but also law enforcement institutions are trapped in a self-reinforcing equilibrium that reduces incentives for those who want to behave honestly and make accountability unlikely (Da Ros and Taylor, 2022).

In addition, many incumbents, mainly elected ones, enjoy privileged jurisdiction, i.e. authorities in certain ranks of public office are entitled to have any criminal cases against them litigated in the higher courts. For example, members of the Brazilian National Congress (Congresso Nacional), ministers, and the president's special jurisdiction is the *Supremo Tribunal Federal* (Supreme Court); governors and state Court of Accounts members, in turn, are trailed by the *Superior Tribunal de Justiça* (Superior Court of Justice). The appointment process for judges in the higher courts is primarily political. Not by chance, the outcomes of the majority of scandals and investigations involving high-level officials are considered disappointing because, after lengthy delays due to judicial loopholes and multiple appeals in courts, they bring few sanctions on the perpetrators. In this respect, the *Mensalão* case needs to be seen as a remarkable exception because, for the first (and only) time, senior politicians were found guilty in a criminal trial and sentenced to prison terms for corruption charges with no reversals or annulments on procedural grounds. *Mensalão* literally means "big monthly allowance," a reference to the periodic illegal payments to congressional representatives to secure their legislative support for Lula's government.

Brazilians call anti-corruption processes with non-conclusive or very few tangible outcomes *"acabou em pizza,"* literally "it ended up in pizza."[6] On the one hand, this sequence of grandiloquent discoveries making headlines in news articles that end up in elite impunity has raised awareness of corruption and explains why corruption is one of the main concerns for Brazilians (France, 2019; Sadek, 2019). On the other hand, this series of scandals does not seem sufficient to explain why petty corruption experienced by citizens affects relatively few people (Ang, 2020; Da Ros and Taylor, 2022) if compared to the overall perception that corruption is widespread.

This puzzling situation is well illustrated by cross-national corruption indices and rankings, which show that Brazil's overall performance has never been dreadful (Power and Taylor, 2011). The 2019 Transparency International's Global Corruption Barometer (TI-GCB) in Latin America and the Caribbean, for example, highlighted that nine out of every ten Brazilians see corruption in the public sector as a major problem but only a small number of people reported paying bribes in their dealings with services, such as schools (4%), hospitals (5%), and the courts (5%). According to the 2019 TI-GCB, 8% of the Brazilian population declared at least one experience with corruption in the previous year. Yuen Yuen Ang's surveys of experts in 15 countries also ranked Brazil as performing well on speed money (also known as grease money) and petty theft (non-elite corruption) but worse than average in terms of access money and grand theft (elite corruption) (Ang, 2020).

For Lucio Picci (2024, p. 124), the statistics on victimisation, with people frequently reporting a personal experience of corruption, support a pessimistic view regarding the topic in Brazil. The 2020 Latinobarometro also illustrates how corruption is perceived as a political issue in the country, positioning office holders and politicians as the most corrupt individuals. When asked to what extent certain categories are corrupt, 11.4% of respondents agreed that all presidents and their advisers are corrupt, 15.8% agreed that all congressional members are corrupt, and 10.7% said that local governments are involved in corruption too. Other categories registered lower figures when the statement was "all of them are corrupt," such as judges (8.5%), entrepreneurs (8.2%), civil servants (6.1%), and tax officials (5.8%). At the same time, over one-third of respondents (36.7%) agreed that corruption had increased a lot over the past year. They also perceived that very little progress had been made in reducing corruption (43.4%).

In 2020, corruption was viewed as one of the six main issues facing Brazilians, behind health, unemployment, education, political situation, and economic problems, according to the 2020 Latinobarometro. In 2018, in the same survey, corruption was the second major issue, behind health, and, in the previous year it was viewed as the main problem in Brazil. Intolerance towards corruption is also high among Brazilians: 67.1% do not think that "it is all right to pay the price of some corruption if problems are solved" (53.5% strongly disagree and 13.6% disagree), although 52.4% considered that the level of corruption in the country had increased over the past year, according to the 2020 Latinobarometro. This survey also indicated how Brazilians are divided about the possibility of eradicating corruption from politics: while 51.6% of respondents think that it is possible, 45.3% disagree.

Action and reaction: accountability institutions, scandals, and anti-corruption solutions

The widespread perception of corruption in Brazil is influenced by the country's long history of corruption scandals and anti-corruption efforts since democracy was restored in the mid-1980s. With the enactment of the Brazilian 1988 Constitution, public agencies, the press, and civil society organisations (CSOs) were assured legal grounds to engage in at least one of the three primary accountability mechanisms: oversight, investigation, and sanction (Carson and Mota Prado, 2014). This "web of accountability" fosters collaboration and competition, mainly among public agencies with a few overlapping roles, as is the case with the Federal Police and the Prosecution Service and with the *Controladoria-Geral da União* (CGU – Office of the Comptroller General) and the *Tribunal de Contas da União* (TCU – Federal Court of Accounts) (Odilla and Rodriguez-Olivari, 2021, p. 127), but also with CSOs (Odilla and Veloso, 2024). Over the years, collaboration and competition among public agencies observed in the "web of accountability" have influenced technological innovation in anti-corruption. Not only has this facilitated government and civil society anti-corruption efforts through the publication and opening of large amounts of public data, but it has also led to some in-house development and improvement of technologies that combat corruption, bolster accountability, improve transparency, and promote integrity, primarily in public expenditures, including procurement, which is a continuing focus of corruption in Brazil.

Although Brazil does not have a single anti-corruption agency, a closer look at the Brazilian horizontal network of institutions of accountability at the federal and regional level suggests that every agency within the public administration could, in principle, conduct investigations and propose administrative sanctions for misconduct, encompassing corruption, via internal affairs units when such units exist. This means that any agency is entitled to deploy its own anti-corruption and pro-integrity mechanisms, including channels to receive reports from civil society and tools to allow civic monitoring. However, there are also law enforcement agencies with distinct formal responsibilities within the executive and legislative branches and many of their responsibilities closely align with the functions expected of an anti-corruption agency.[7]

There are certain limitations in terms of imposing penalties and bringing cases to court. For example, at the federal level, the TCU and the CGU can impose certain types of sanctions but cannot bring cases to criminal courts. While the Federal Police and the *Comissão Parlamentar de Inquérito* (Parliamentary Committee of Inquiry) are entitled to conduct investigations and indict individuals for corruption, the Prosecution Service, which can also conduct investigative procedures, is the sole governmental agency that is allowed to bring charges and carry out anti-corruption criminal and civil procedures in court. Additionally, the judiciary is the ultimate sanctioning authority, as it can review and rescind punishments imposed by administrative law enforcement bodies (Odilla and Rodriguez-Olivari, 2021, p. 121).

Primarily due to the actions of this "web of accountability" institutions (Mainwaring and Welna, 2003), mainly to monitor, investigate, and sanction political wrongdoings, corruption scandals became a common feature in Brazil. After democracy was restored in 1985, every Brazilian president who has taken office has had to contend with scandals that have tainted their administration. In response they have promoted anti-corruption reforms, including passing new legislation and creating new agencies. Many of the changes are related to institutional arrangements aligned with the increasing digitalisation at that particular time. President José Sarney (1985–1990), for example, faced a congressional inquiry that found endemic corruption in his administration and recommended his impeachment, which was later vetoed. Amid scandals and investigations, Sarney's administration created the *Sistema Integrado de Administração Financeira do Governo Federal* (SIAFI – Integrated System of Financial Administration) that centralised and digitalised budget data, which over the years proved to be essential in making fiscal data available for oversight purposes (Lopes, 2018).

President Fernando Collor de Mello (1990–1992), who during the 1989 campaign was nicknamed the "hunter of maharajahs" due to his penchant for attacking the high salaries and perks of public officials (Figueiredo, 2010), saw his popularity wane as scandals involving the embezzlement of public funds in his administration accumulated. Before he resigned in a failed attempt to avoid impeachment (Figueiredo, 2010; Power and Taylor, 2011), his administration enacted a series of anti-corruption Acts that served as the basis for the work of anti-corruption agencies at the federal level. Among them are the Public Servants Act (Law No. 8112/1990), which applies sanctions to bureaucratic misconduct ranging from absenteeism to bribery; the Improbity Act (Law No. 8429/1992), which punishes illicit personal enrichment, damage to the public budget, and violation of administrative principles (Mota Prado and Cornelius, 2020, p. 13), and Law No. 8443/1992, which regulates the TCU.

Shortly after the aforementioned *Anões do Orçamento* scandal came to public notice, Collors' vice-president Itamar Franco (1992–1994) granted new resources and attributions to the *Receita Federal* (Federal Revenue Service) to conduct more effective oversight of financial transactions (Lagunes et al., 2021a). Franco also enacted Law No. 8666/1993, which lays down the general rules about public contracts and public bids, making them more transparent and accountable (Avritzer and Filgueiras, 2011). This was the starting point for digitalising the public administration through Decree No. 1048/1994 to organise what at that time were known as "systems of information and informatics."[8]

In 2000, Brazil introduced innovative electronic bids and initiated the digitalisation of public procurement procedures through Decree No. 3697/2000. Coincidentally, this occurred just a few months before the son of President Fernando Henrique Cardoso (1995–2002) came under investigation over allegations of corruption. He was suspected of organising an expensive exhibition stand at an event in Germany to promote the 500th anniversary of the arrival of the Portuguese in Brazil (Folha de S.Paulo, 2000), which raised questions about the public procurement criteria used during his father's administration.

Under Cardoso's administration, other anti-corruption control mechanisms were also rapidly implemented amid scandals. Accusations of corruption involving his administration and coalition members prompted multiple calls to launch congressional inquiry committees (Odilla and Rodriguez-Olivari, 2021). Two scandals in particular came to light through reports published by the *Folha de S.Paulo* newspaper:

first, the supposed buying of congressional votes to secure the approval of a legislative amendment that would allow Cardoso and future presidents to run for re-election (Rodrigues, 2014), and second, the release of audio files suggesting direct interference by the federal government in the privatisation of telecommunication companies (Rodrigues and Lobato, 1999). The latter scandal led to the resignation of the minister of communications, the president of the *Banco Nacional de Desenvolvimento Econômico e Social* (BNDES – National Bank for Economic and Social Development), and executives of the *Banco do Brasil* and Previ, the pension fund for bank employees, who were allegedly at the centre of the supposed schemes (Folha de S.Paulo, 2003). Cardoso's administration also saw a scheme of extensive collusion between congressional members, their relatives, and employees of the development agencies for the Amazon and the northeastern regions being spotted by many press investigations (Odilla and Rodriguez-Olivari, 2021).

Cardoso's major response came in 2001, the year before the presidential election, when he created an internal affairs department to investigate and punish civil servants more rapidly: the *Corregedoria Geral da União* (CGU – Office of the Inspector General) which two years later became the *Controladoria Geral da União*, keeping the acronym CGU despite the extension of its responsibilities) (Odilla, 2020a, 2020b). In 2000, he enacted the Fiscal Responsibility Act (Complementary Law No. 101/2000), which was considered the benchmark for governmental transparency in Brazil because it established clear sanctions for those who do not respect the budgetary and financial guidelines at the three levels of government (Castro Neves, 2013).

Table 2.1 gives the names of Brazil's presidents, the most significant corruption scandal during the period (not necessarily linked to the president), and the anti-corruption legislation passed during each presidential administration.

During Cardoso's administration, an investigation called Banestado made public a case of tax evasion, tax avoidance and money laundering through offshore money flows involving politicians, businesspeople, bankers, and illegal foreign exchange traders (*doleiros*). This scandal prompted the creation of an anti-money laundering law and the Council for Control of Financial Activities. The new agency was both a response to treaty commitments and a means to staunch capital outflows that were undermining the government's fixed exchange rate strategy (Praça and Taylor, 2014, p. 34). In 1999, the *Comissão de Ética Pública* (Public Ethics Commission) was created as a consultive body charged with overseeing integrity. Over the years, all these new governmental agencies started generating and storing (increasingly

Table 2.1 Major scandals and key administrative and legal anti-corruption reforms in Brazil, 1985–2023

Years	Presidential administration	Biggest corruption scandal(s) to come to public notice	Relevant anti-corruption legislation
1985–1990	José Sarney	Ferrovia Norte-Sul (Northern-Southern Railroad, 1987); CPI da Corrupção (Congressional Inquiry on Corruption, 1988); Caso BR/Petrobras (1988)	Creation of the Integrated System of Financial Administration (SIAFI) (Decree No. 95519/1987); Public Civil Action Law (Law No. 7347/1985); Money Evasion Law (7492/1986); Public Bidding Decree (Decree No. 2300/1986)
1990–1992	Fernando Collor de Mello	Jorgina de Freitas/INSS (Fraud in the Social Benefit Service, 1991); Rosane Collor (1991); PC Farias/Collorgate (1992)	Public Servants Act (Law 8112/1990); Improbity Act (Law No. 8429/1992); new regulation regarding the Court of Accounts (Law No. 8443/1992); electoral ineligibility (Complementary Law No. 64/1990); Archives Law to preserve public documents (8159/1991); Taxpayer Identification Law (Law No. 8021/1990)
1992–1994	Itamar Franco	Rodomar and the BNDES (1992); Anões do Orçamento (Budget Dwarves, 1993)	Law of Bidding and Tendering (8666/1993); creation of the System of Information and Informatics Resources (Decree No. 1048/1994); new normative response for preventing and suppressing violations of the economic order (Law No. 8884/1994); Ethical Code for Civil Servants in the Federal Executive (Decree No. 1171/1994); Elections Law to limit business campaign contributions (Law No. 8713/1993)

Years	Presidential administration	Biggest corruption scandal(s) to come to public notice	Relevant anti-corruption legislation
1995–2002	Fernando Henrique Cardoso	TRT de São Paulo (São Paulo Labour Court, 1999); Banestado money laundering (1990s); Pasta Rosa (Pink Folder, 1995); SIVAM Amazon-surveillance project (1995); vote buying for the re-election amendment (1997); privatisation of telecoms/wireless (1998); Bank Marka and FonteCindam (1999); Precatórios do DNER (Judicial Debts of DNER, 1999); Sudam/Sudene embezzlement scheme (2000)	New rules for transparency of electronic voting (Law No. 10408/2002); new limits for civil service expenditure (Complementary Law No. 82/1995); creation of the Council for Control of Financial Activities and anti-money laundering law (Law No. 9613/1998); creation of the Public Ethics Commission (Decree with no number/1999); Code of Ethics and Conduct for Higher Ranks in the Federal Executive, Presidency and Vice-Presidency (Decree with no number/1999 and Decree No. 4081/2002); new forms of public bidding in procurement (Law No. 10520/2002) Fiscal Responsibility Act (Complementary Law No. 101/2000); enactment of the OECD Anti-Bribery Convention (officially the Convention on Combating Bribery of Foreign Public Officials in International Business Transactions); Criminalisation of Vote Buying (Law No. 9840/1999); creation of the Office of the Inspector General (Provisional Measure 2143–32/2001 and Decree No. 4177/2002);

Years	Presidential administration	Biggest corruption scandal(s) to come to public notice	Relevant anti-corruption legislation
			organisation and regulation of the Federal Planning and Budgeting System, the Federal Financial Management System, the Federal Accounting System, and the Internal Control System of the Federal Executive Branch (Law No. 10180/2001); regulation of hearings and meetings granted to private individuals by public agents (Decree No. 4334/2002); enactment of the Inter-American Convention against Corruption (Decree No. 4410/2002)
2003–2010	Luiz Inácio Lula da Silva	Operation Anaconda (on judicial corruption, 2003); Waldomiro Diniz (2004); Máfia dos Vampiros (Vampire Mafia, 2004); Mensalão (Big Monthly Allowance, 2005); Sanguessugas (Leeches, 2006); Palocci Case (2006); Boi Barrica/Faktor (2006); Dossiê dos Aloprados (Madmen Dossier, 2006); Operation Navalha (2007); Mensalão Tucano (Big Monthly Allowance of the PSDB party, 2007); Operation Hurricane (2007); Satyagraha (2008);	Creation of the Transparency Portal (2004); transparency law for budget transparency and reporting (Complementary Law No. 131/2009); System of Management of Agreements and Transfer Contracts Agreements (Sincov) (Decree No. 6428/2008) Change the Code of Ethics and Conduct of Civil Servants (Decree No. 4610/2003); Criminal Code Reform to enhance penalties for bribery (Law No. 10763/2003); creation of the Office of the Comptroller General (Law No. 10683/2003); creation of the Public Transparency and Anti-Corruption Council (Decree No. 4923/2003);

Years	Presidential administration	Biggest corruption scandal(s) to come to public notice	Relevant anti-corruption legislation
		Cartões Corporativos (Corporate cards, 2008); Mensalão do DEM (2009); Castelo de Areia (Sand Castle, 2009); Atos Secretos (Senate's Secret Acts, 2009); Erenice Guerra Case (2010)	creation of the Internal Affairs System for the Federal Executive (Decree No. 5481/2005 and Decree No. 6692/2008); changes in the structure of the CGU, creating the Secretariat for Corruption Prevention and Strategic Information (Decree No. 5683/2006; enactment of the United Nations Convention against Corruption (Decree No. 5687/2006); creation of the Ethics Management System of the Federal Executive (Decree No. 6029/2007); creation of oversight bodies for the judiciary (*Conselho Nacional de Justiça*, CNJ) and the prosecution service (*Conselho Nacional do Ministério Público*, CNMP) (Constitutional Amendment No. 45/2004; Ficha Limpa Law (Clean Slate Law, Complementary Law No. 35/2010)
2011–2016	Dilma Rousseff	Palocci (2011); Conab (2011); Ministry of Labour and NGOs (2011); Segundo Tempo (Second Half, 2011); Máfia dos Transportes (Transport Mafia, 2011); Cachoeira/Monte Carlo (2012); Lava Jato (Car Wash, 2014); Zelotes (2015)	Access to Information Law (Law No. 12527/2011); Anti-Corruption Act (Clean Companies Law, Law No. 12846/2013); Organised Crime Control Act with new rules for plea bargaining (12850/2013); Differentiated Public Procurement Regime Law (Law No. 12462/2011); Brazilian System for Defense Competition (12529/2011); new Anti-Money Laundering Law (12683/2012); Conflict of Interest Law (Law No. 12813/2013)

Years	Presidential administration	Biggest corruption scandal(s) to come to public notice	Relevant anti-corruption legislation
2016–2018*	Michel Temer	Greenfield (2016); JBS (2017); Candidaturas laranjas (Orange Candidates, 2018)	State-owned companies' liability law to reduce political appointments (Law No. 13303/2016); General Personal Data Protection Law that has been used to deny access to public information (13709/2018)
2019–2022*	Jair Bolsonaro	Orçamento Secreto (Secret Budget, 2021); Caso Queiroz/Rachadinha (Queiroz Affair, 2019); Laranjal do PSL (PSL Orange Grove, 2019); Escândalos das Joias (Jewellery Scandal, 2022); Escândalo do MEC (Minister of Education Scandal, 2022); Fraudulent Codevasf contracts (2022); Covaxin scandal (2022)	New Administrative Improbity Law that loosens punishments (Law 14230/2021); new Abuse of Authority Act, which can be used to intimidate public officials involved in cases concerning powerful individuals (Law No. 13869/2019); reform of the Money Laundering Act, weakening the administrative structure of the Financial Activities Control Board; changes in the Penal Code to introduce whistleblowing protection (Law No. 13964/2019 and Decree No. 10153/2019); Federal Government System for Managing Partnerships (Decree No. 11271/2022), with the launch of the e-platform Transferegov (former Plataforma+Brasil)
2023–	Luiz Inácio Lula da Silva	Football Match-Fixing, (2023); Aviões da FAB (misuse of Brazilian Air Force's flights by authorities, 2023)	General Law for Sports (Law No. 14597/2023), with the criminalisation of private corruption in sport and other measures to combat the manipulation of sporting results

Source: based on Taylor and Buranelli (2007); Power and Taylor (2011); Carson and Mota Prado (2014); Avritzer and Filgueiras (2011, pp. 37–40); Freire (2015); Chemim (2017); Lagunes et al. (2021a, 2021b); Da Ros and Taylor (2022, pp. 73–75, 225–234).

Note: *During the presidencies of Michel Temer and Jair Bolsonaro many setbacks in Brazil's legal and institutional anti-corruption frameworks were observed (France, 2019; Vieira and Miranda, forthcoming).

digitalised) data related to suspicious financial transactions, disclosures by public officials of their finances and assets, intergovernmental financial transactions, and public procurement procedures, among others, that proved crucial to fighting corruption.

A popular initiative criminalising vote buying in electoral campaigns was sanctioned by President Cardoso in 1999. However, the vote-buying law applies only to candidates running for elections. It does not cover, for example, side payments made by the executive to congressional members to pass bills of interest. This was the essence of the aforementioned *Mensalão*, the first major scandal that threatened to bring down President Luiz Inácio Lula da Silva (2003–2010).[9] In the wake of *Mensalão*, representatives from Lula's Workers' Party and opposition parties faced investigations and criminal trials for similar quid pro quo schemes (Lagunes et al., 2021a). Even amid persisting scandals that made headlines, important anti-corruption initiatives were offered greater support during Lula's administration than in the previous ones. This could be seen in the increasing number of anti-corruption investigations and audits conducted by the federal government and a greater number of parliamentary inquiries in Congress.

In 2004, one of these anti-corruption audits to inspect municipal finances discovered that overpriced ambulances were being bought with public money. The scandal became known by the name given by the Federal Police to the operation, which was launched in 2006. *Sanguessugas* means leech, a parasitic worm that feeds by attaching itself to its mostly unsuspecting hosts and sucking their blood, just as the politicians and businesspeople involved in the scandal appeared to be doing to the health sector (Petherick, 2015). A special congressional inquiry committee recommended that both the Senate and Lower House open disciplinary procedures against 72 members of Congress (CPMI, 2006). None of the representatives were sanctioned by their peers, but many did not run in the next election. Only five out of 69 members of the Lower House and one out of three senators managed to get re-elected in the next election. As a result of the scandal, Congress implemented stricter rules and imposed greater transparency on congressional amendments (Da Ros and Taylor, 2022).

Opening data to promote transparency

The anti-corruption agency responsible for the audits that identified the "bloodsuckers" was the CGU. During Lula's administration, the CGU became an anti-corruption agency that, apart from its original internal affair responsibilities, implemented new initiatives, proposed bills, and enacted regulations aimed at curbing corruption at the federal level (Odilla and Rodriguez-Olivari, 2021). The CGU was the first governmental agency to promote active downward transparency in Brazil. In 2004, it launched a *Portal da Transparência* (transparency portal) with information on budget and spending and then started to develop significant guidelines to be followed by other branches at the federal, state, and local level. However, the CGU has no power to guarantee transparency in all the Brazilian public bodies, because its jurisdiction is limited to the federal executive. This helps to explain why advancements and the implementation of a transparency infrastructure are incremental but non-linear across the country.

In some cases, the pressure of the media was crucial to compel other branches to provide open data. For example, it was not until February 2009, after several scandals involving the misuse of public money, that the Lower House decided to provide active transparency of congresspeople's expenses while they are in office. Each representative is entitled to reimbursement for expenses related to meals, travel, office stationery, and phone calls. Yet the data were limited to overall figures, invoice numbers, and the names of suppliers paid with public money. In August 2009, the Supreme Court ruled in favour of the newspaper *Folha de S. Paulo* and required the Lower House to present the receipts.[10] Under pressure, the Lower House ruled that representatives must upload all receipts refunded, and micro-level data on their expenditures are now published on the congressional web portal. This made it possible years later for civil society's digital initiatives to hold these expenditures accountable, as will be seen in the next chapters. In 2009, Congress also approved Complementary Law No. 131, dated May 27, 2009, known as the Transparency Law or Capiberibe Law, which requires the real-time disclosure of revenues and public expenditures of the federal government, states, and municipalities on the internet.

From 2003 to 2016, during the Workers' Party time in office, there were other notable advances in the anti-corruption agenda. In 2003, the Federal Police carried out Operation Anaconda, breaking up a syndicate of lawyers, detectives, and corrupt judges (Taylor and Buranelli, 2007). As a result of the scandal, judicial reform was carried out in 2004; two administrative oversight bodies were created to provide guidance and to investigate and punish officials and members of the judiciary (CNJ) and the prosecution service (CNMP) in violation of the law or ethical rules (Carson and Mota Prado, 2014). One of the first decisions of the CNJ was to prohibit nepotism in the judiciary. In 2008, the Supreme Court issued a decision extending the prohibition to all three branches of the government, making the hiring of relatives who have not passed a civil service entrance examination a violation of the Constitution in the executive and legislative branches at the federal, state and local level. Yet, the oversight function played by the CNJ and the CNMP has attracted criticism for often protecting members of the judicial system instead of holding them accountable (Oliveira, 2017).

Popular pressure to clean up politics

Drafted when Brazil had returned to civilian rule after two decades of military dictatorship (1964–1985), the 1988 Constitution laid the foundations for the country's web of accountability institutions in which governmental agencies, civil society, and the press devote attention to the monitoring, investigation, and sanctioning of corruption (Power and Taylor, 2011; Carson and Mota Prado, 2014). The new Constitution also became known as the "citizens' Constitution" because it introduced mechanisms to increase citizens' participation in the public sphere, including law-making. As Mattoni and Odilla (2021, pp. 1133–1134) noted, since 1988, four bills converted into laws have been recognised as successful popular initiatives. Two of them are related to anti-corruption mechanisms: the 1999 Anti-Vote Buying Law classifies the practice as an electoral infraction, and the 2010 Clean Slate Act, or *Ficha Limpa*, imposed an eight-year ban that prevented candidates convicted by a collegiate for certain types of crimes, including corruption, from running for election.

In 2010, Lula also enacted *Ficha Limpa*, which, ironically, was used to disqualify his candidacy eight years later when he registered to run in the presidential election while still in jail. The *Ficha Limpa* campaign was led from 2007–2010 by the Movement to Combat Electoral Corruption, a group representing more than 40 CSOs, non-profits, and religious associations. These included the National Confederation of the Bishops of Brazil and the Justice and Peace Commission, at that time headed by the social activist Francisco Whitaker, one of the founders of the World Social Forum (Mattoni and Odilla, 2021, p. 1135). *Ficha Limpa* was signed into law in June 2010, just before a general election and nine months after the submission to Congress of the bill signed by 1.6 million people. This was a noticeably short time frame for the Brazilian Congress, which on average takes 45 months to approve this type of legislative proposal (JOTA, 2019; Mattoni and Odilla, 2021). The legislation has been effectively applied to ban candidates from running for office: 185 were banned the in 2022 presidential race and 173 in 2018 (Mendes, 2022).

In 2011, President Dilma Rousseff (2011–2016) succeeded Lula under popular pressure to tackle corruption and began her term by promoting what became known as *faxina ética* (ethical cleaning). In her first months in office, she fired six ministers suspected of corruption whose misconduct was making the headlines (Lagunes et al., 2021a). Under Rousseff's administration, the Access to Information Law was enacted in 2011. The bill, which had been introduced in Congress in 2009, was approved, with high pressure coming from CSOs and the support of the CGU (Lagunes et al., 2021a; Odilla and Rodriguez-Olivari, 2021). The Access to Information Law became a powerful tool for journalists and activists to access and analyse data and compile the information for general release. In addition, the law explicitly states that access to data should be granted in electronic and open formats if so requested, which has made it easier to launch many digital anti-corruption initiatives.

Despite the apparently vigorous way Rousseff began her presidency, by 2013, crowds of protesters had taken to the streets to express their dissatisfaction with the quality of public services, especially transportation, health, and public security, and with the billions spent on financing major sports events such as the 2014 World Football Cup and the 2016 Rio Olympics (Gohn, 2014). To counter public rage, Congress approved and Rousseff enacted two legislative bills to regulate plea bargaining for criminal law proceedings and create leniency agreements with companies for administrative law proceedings (Lagunes et al., 2021a). Both laws were largely used to push the *Lava Jato* investigation forward.

The scandals emerging from the *Lava Jato* operation also promoted immediate changes in the campaign financing rules. Over the years, Brazil has improved the electoral rules, obliging candidates to provide a declaration of assets and income and forbidding certain types of expenditure such as musical concerts for promoting candidates. Electoral courts also had data on political financing and spending made more transparent and available online relatively systematically (Da Ros and Taylor, 2022). After *Lava Jato* became public, though, Brazil banned corporate donations to parties and candidates. The outcomes of this reform remain controversial, as companies still influence elections and wealthy candidates benefit from the new rules since individuals are allowed to contribute up to 10% of their last annual declared income (Benjamin and Couto, 2016).

The mega-investigation also triggered a rise in public anger and increased concerns about corruption and the perception that corruption is one of the worst problems facing the country (Picci, 2024). If the 2013 protests were sparked by a very diffuse agenda, the *Lava Jato* probe contributed to a new wave of street demonstrations, also organised online but more focused on the anti-corruption discourse and against the Workers' Party administrations. In 2015 and 2016, protesters returned to the streets to demonstrate their support for the law enforcement agents directly involved in the investigations and against the corruption scandals, mainly those involving the Workers' Party itself and its supporters. They demanded the impeachment of then-President Dilma Rousseff, who was ousted from office in 2016 on charges of manipulating the federal budget (Odilla, 2024).

Rousseff's vice-president, Michel Temer (2016–2018), took office but did not remain untouched by corruption scandals. He lost ministers to corruption investigations and faced scandals of graft. He was wiretapped by one of the country's most important businesspeople allegedly approving a bribe, which resulted in Temer being jailed for four days, three months after leaving office, as part of a massive corruption investigation. He was accused of receiving bribes and attempting to buy silence to avoid a plea bargain with the former speaker of the Lower House, who was also in jail under the scope of *Lava Jato*. Following Jair Bolsonaro's election to the presidency in 2018, his administration frequently broke his electoral promises to fight corruption (Lagunes et al., 2021a). First, a minister became a defendant in a scheme of vote buying during the elections. Second, a government official who reportedly asked for a vaccine deal bribe faced graft accusations. Finally, a system of opaque yet legal budgetary grants was created to maintain a loyal support base in Congress suspected of illegally expending the money (Rezende, 2021; Pires, 2022).

During both Temer's and Bolsonaro's presidencies, scandals kept emerging, but the country started to encounter bigger difficulties in advancing wider reforms and facing institutional setbacks in its accountability framework (France, 2019). These include the weakening of anti-corruption legislation, growing interference in law enforcement agencies, overruling of court decisions with systemic impact, and growing issues with active and passive transparency, including data blackouts of official figures such as COVID-19-related infections and deaths. When Lula was re-elected to the presidency in 2022, doubts lingered about whether he would prevent setbacks and strengthen the anti-corruption system that he helped to build. In a way, though, the ensuing investigations were used to expose wrongdoings during his previous terms that ultimately led to his imprisonment.

On the one hand, the high perception that corruption is a major problem over the years may help to explain why the country has seen many bottom-up initiatives emerge since the late 1990s to fight corruption, many of them deploying digital technologies. Where corruption is perceived as a resilient major issue, as is the case of Brazil, a natural demand for action or reaction is believed to exist. As many of our research participants pointed out, ordinary citizens started to ask what they could do with their knowledge and expertise to help to curb corruption, and techy-savvy civil servants working in law enforcement agencies were encouraged to use their skills to improve outcomes. On the other hand, a relatively high level of scepticism related to anti-corruption efforts may also explain the often small scale and scope of these emerging digital initiatives and their overall low-level engagement in terms of numbers, including the use of their digital anti-corruption tools. Before exploring how digital media has been developed and used to fight corruption, it is important to understand Brazil's path toward digitalisation. Next, the discussion will look at how Brazil advanced its digital system to improve its control capacities to fight corruption and maladministration.

Digitalising anti-corruption efforts

Although the internet and mobile telephony arrived in the country in 1988 and in 1990, respectively (Pedrozo, 2013), it was not until 1994 that the country took its first important step towards digitalisation with the creation of the *Sistema de Administração dos Recursos de Tecnologia da Informação* (SISP – Brazilian System for the Administration of Information Technologies Resources). The SISP was established by decree in 1994 and updated in 2011. It provided institutional guidelines to organise, control, and supervise the information technology resources of the Brazilian executive at the federal level.

A review published by the Organisation for Economic Co-operation and Development (OECD) in 2020 of Brazil's digital governance emphasises that the country did not take a more consistent path towards digitalisation until 2000. However, processes of large-scale digitalisation in banking and public finance, including taxation and public procurement, started earlier and can be seen as the initial techno-infrastructural mainframes that later facilitated the process of digitalising anti-corruption. Elections in Brazil have also been the subject of early digitisation to curb corruption. First tested in the 1996 elections, the electronic ballots aimed to eliminate fraud in the electoral process by removing the human hand from the counting process (McCormack, 2016; TSE, 2014).[11] In 2000, electronic ballots became the sole voting method in Brazil, making it the first country in the world to conduct elections entirely through an electronic voting system (Neves and Silva, 2023).

The electronic voting system was just one of many governmental digital initiatives that emerged in the 1990s. The Brazilian *ComprasNet* website was launched in 1997 only to call for tenders and publish summaries of signed contracts by the federal executive. In the 2000s, it became an e-procurement platform to increase competitiveness and social control over public spending (BNDES, 2002). *ComprasNet*, which later became the purchase portal of the federal government, is currently a key database for many anti-corruption technologies in Brazil (Odilla, 2023a).

Following this logic to increase control over public spending in the country, in 2004 the CGU, as mentioned earlier, created its transparency portal. It became one of the most important initiatives of e-government and active transparency in the country due to the amount of open data available. Initially, the portal aimed to advance fiscal transparency through open government budget data and, therefore, information that was already kept in the SIAFI became publicly available. Internal tensions grew regarding the opening of fiscal information related to federal transfers to local governments and information on funds directly received from public resources (Graft et al., 2016). In the end, active transparency prevailed, and the portal expanded its content. Its scope has broadened to an easily accessible website that stimulates citizen engagement, and many local governments and other Latin American countries have used it as a basis for similar online tools.

In 2005, Brazil took another important step by approving inter-operability standards of e-government (*Padrões de Interoperabilidade de Governo Eletrônico* – e-PING) to improve the connectivity between the information systems of the different sectors of government, considered crucial to advance its evolution towards digital government. In the following years, Brazil kept improving its legal frameworks for digitalising the government that often involved more open and collaborative processes with civil society (OECD, 2020). Hence, it can be seen as a good example of cultivating engagement with citizens when developing strategies for digital transformation.[12]

Several initiatives illustrate this culture of openness and engagement. The *Marco Civil da Internet* (Brazilian Internet Bill of Rights), for example, was the result of wide public consultation before becoming law (Law No. 12965/2014), addressing rights and duties related to topics such as freedom of expression, privacy, and personal data management, and net neutrality (OECD, 2020). The Brazilian Strategy for AI,[13] a 2021 document outlining ethics and governance guidelines for AI systems prepared by the Brazilian government, was also based on over 1,000 contributions collected through public consultation. Another example of efforts related to the inclusiveness of the design and implementation of digital policies is the Brazilian Action Plan for Open Government, which was developed through workshops with civil society representatives and public officials. Not by chance, Brazil was one of the eight founder countries that in 2011 endorsed the Open Government Partnership, which currently includes over 70 countries; Brazil also had a prominent role in its development (Brelàz et al., 2021). Through open government, this multilateral initiative aims to harness technology to strengthen governance, empower citizens, and fight corruption.

Harnessing digital data and platforms for bridging gaps

In the wake of openness and inclusiveness, Brazil started to build unified solutions and bridge information silos. To facilitate requests and access to information under the new legislation, the CGU began to offer training, support, and booklets for government bodies in the federal sphere, states, and municipalities. It likewise offered the source code of the *Sistema Eletrônico do Serviço de Informação ao Cidadão* (e-SIC – Electronic Citizen Information System), through which anyone can make requests and appeals and receive information online (Ministério da Transparência e Controladoria-Geral da União, 2018). Created in 2011, the e-SIC became a remarkable example of passive transparency (Castro Neves, 2013). The system was replaced in 2019 by another platform (Fala.BR) that unified requests for information and manifestations to the ombudsman (Odilla and Rodriguez-Olivari, 2021).

However, the Access to Information Law also made public issues related to managing paper-based records, which forced the administration to speed up the electronic management of public documents and procedures (Saraiva, 2018). The *Sistema Eletrônico de Informação* (SEI, Electronic System of Information) is a solution developed by the judiciary in southern Brazil that granted it free of charge to other public agencies. It is envisaged that the system will eventually be adopted by the entire federal public administration to replace paperwork. The expansion process of SEI was initially led by the Ministry of Planning and is one of the many cases in which tech-savvy governmental workers develop an internal solution that gains scale and becomes adopted more broadly in the public administration.[14] More recently, Brazil launched a centralised portal for citizens (Portal Gov. br) with a unique digital authentication system to unify digital portals and provide simplified methods for finding and using available public services. In June 2023, 4,200 services were available from over 200 governmental agencies (Governo do Brasil, 2023).

Once public administration, at least at the federal level, intensified its digitalisation process, a series of platforms to promote data-sharing practices became available online. For example, in December 2011, the Brazilian Portal of Open Data (dados.gov.br/) was launched to offer a data catalogue with searching mechanisms, metadata, and clear license information and, of course, access to the datasets themselves, similar to those made available in countries such as Canada and the United Kingdom (Castro Neves, 2013). Later, other initiatives were implemented such as Conecta.GOV, which provides a catalogue of the federal government application programming interfaces, and the federal platform for data analytics and data-driven policymaking and analysis. By opening and releasing these datasets in machine-readable formats, the Brazilian government aims to follow their guidelines, but joint efforts by civil servants from different agencies, CSOs, and the media are crucial in the actual co-creation of public value beyond data availability and accessibility.

This chapter has focused on how scandals have driven the digitalisation of anti-corruption and showed how many government measures have been taken in response to them. But it has to be said that the digitalisation of anti-corruption was facilitated by an earlier and longer process of digitalisation of public administration. The multinational technology company IBM, for example, began operating in Brazil in 1917 under the name of Computing Tabulating Recording Co (CTR) to organise and carry out the census three years later. Despite IBM's longstanding presence in the country, Brazil remained mostly closed to foreign investment for decades, with restrictions on access to technology and the operation of foreign companies affecting the modernisation of public administration. Despite this, the country can be considered an early adopter of digitalisation in Latin America. The first credit card in Brazil dates to 1956 (Gomes and Costa, 2015), and since then the digitalisation of payments has expanded rapidly. The government company for data processing services (Serpro) was created by a federal law in 1964. However, it was not until the 1990s that the digitalisation (and digitisation) of the public administration was implemented more consistently. This process was driven by the 1990 administrative reform, influenced by the new public management agenda to improve efficiency, effectiveness, transparency, public management control and the need for accountability (Bresser-Pereira, 1998; Gaetani, 2005; Cristóvam et al., 2020).

Conclusion

As this chapter has shown, in recent decades the public sector in Brazil at the federal level has made substantial progress in integrating digital technologies that have improved the agility of its internal processes. The country created a digital environment that has facilitated, for example, the monitoring of public spending and the cross-checking of large amounts of data. The existence of many loopholes and a lack of preventive checks became clear during the COVID-19 crisis, for example, when social benefit fraud was pulled off by individuals who exploited the weakness of an as yet still not entirely integrated system. In Brazil, emergency aid (*auxílio emergencial*) that could be accessed and released using a dedicated mobile phone application was made available to low-income individuals in situations of informality and whose family income was heavily impacted by the health crisis, emergency aid (Senne, 2021). Nevertheless, more than BR $809 million (around US $160 million) was paid illegally to at least 1.8 million people. According to a report issued by the CGU, illegal payments were made to people who were already dead, who had accumulated different benefits, or who were not unemployed, including cases of political appointees working for local legislatures (Portela, 2022). The emergency aid scandal can be added to the long list of corruption scandals in Brazil.

It must be said that the COVID-19 pandemic created incentives for corruption and fraud in many countries (Rose-Ackerman, 2021). In addition, as Lagunes et al. (2021a) stressed, witnessing scandals during any given administration is not always a synonym for a lack of accountability. It could, instead, reflect the opposite, as the cases of fraud associated with the payment of COVID-19-related social benefit suggest. Administrations more inclined to confront corruption by creating specialised bodies to uncover official graft, enacting new laws to promote integrity and increase penalties, and developing new digital tools to facilitate the oversight and investigative mechanisms may witness an increasing number of scandals over a given period. In the case of Brazil, however, many advances in the anti-corruption legal and institutional frameworks, including some related to digital openness and transparency, have coincided with massive scandals that have troubled incumbent administrations as well as exerting domestic and international pressure.

The 1990s was a decade in which corruption scandals came under the spotlight in Brazil and many other Latin American countries and in which international transparency and accountability standards and guidelines were developed (Pereyra, 2019). Although these two closely related processes followed different political dynamics and motivations, the implementation of policies related to promoting transparency, accountability, and social participation, especially those that improve legal frameworks and use digitalisation and digital media, has been seen by many academics and practitioners as the best response to the growing number of scandals.

Indeed, in the anti-corruption field, enthusiasm is still great, and expectations are high that digital media will offer new effective means for the detection, prosecution, and prevention of corruption, despite the mixed existing evidence and the risks of using technology to engage in corruption (Adam and Fazekas, 2021; Köbis et al., 2022). On the one hand, anti-corruption digital solutions can speed up procedures, make rapid analyses and predictions with large datasets that no human could make as fast as machines, remove certain face-to-face interactions necessary for corruption, and ease awareness and communication due to their lower costs. However, emerging technologies require digital literacy, specific skills and resources to develop and use (Charoensukmongkol and Moqbel, 2014; Odilla, 2024). On the other hand, deploying them does not automatically translate into less corruption. Indeed, these technologies can provide new opportunities for corruption, from spreading false information to forging datasets, hacking electronic systems, and laundering money with cryptocurrencies, to mention just a few (Adam and Fazekas, 2021; Köbis et al., 2022). In addition, anti-corruption efforts can also backfire when "too much accountability" generates setbacks, as was the case in Brazil with its legal and institutional anti-corruption frameworks after the *Lava Jato* (France, 2019; Vieira and Miranda, forthcoming). This is more likely to happen where corruption is systemic and is viewed as a way to keep governments working (Picci, 2024).

Yet despite the public efforts and pressures to upgrade governments' capacity to promote integrity by opening data in the digital age, digital transformation and the use of emerging technologies to fight corruption is not uniform, nor has it made corruption less constant in Brazil. The digitalisation of anti-corruption is far more advanced at the federal executive level than it is at the legislative level. The judiciary branch at the federal level and most state and local governments still struggle to align their practices and procedures with what is expected of an open and digital government.

The digital divide is also a challenge within the Brazilian public administration, where levels of digitalisation, openness, and inclusiveness vary dramatically, mainly due to a lack of capacity, skills, and resources. The inequality in access to basic requirements, such as electricity and internet connection, should also be taken into consideration (Adam and Fazekas, 2021, p. 5), which proved to be the case in Brazil. In April 2022, the Brazilian media revealed that the Bolsonaro administration, driven by "pork barrel" politics, allocated BR $26 million to buy robotic toolkits for schools. However, these schools were grappling with a shortage of computers and internet access; indeed, one school located in a small town in one of the poorest states in the country even lacked access to a water supply (Saldanha, 2022).

The OECD, which has been closely assessing the Brazilian digital transformation, has already noted the urgent need to improve the transparency and accountability of the public purchase of information and communication technologies in Brazil. Its evaluation pointed out issues related to the maze-like legislative framework combined with the absence of an authority coordinating and leading processes to "promote collaboration between different sectors and levels of government, foster co-creation of public value with civil society" and create "cultures of sharing and reuse of government data" (OECD, 2020, chap. 5). In addition, in March 2024, Brazil still lagged behind in its provisions on AI regulation and governance, despite having already deployed AI-based tools, including bots to fight corruption (Odilla, 2023a), and created innovation labs within the field of public administration to test emerging technologies. Unfortunately, the Brazilian AI Strategy launched in 2021 seems more like a protocol of intent than an actual plan of action, as noted in an assessment conducted by the Federal Court of Accounts (TCU, 2022).

All these gaps impact the attempts to bridge digital information and infrastructure and foster a digital government, letting citizens communicate their own needs to co-design more proactive digital services and policies open and auditable by default (OECD, 2020). Critical legal loopholes remain in this regard, considering both the anti-corruption and the digital transformation agendas. Even under these circumstances, the Brazilian network of accountability has improved its capacity, and it is observed a digitalisation of anti-corruption often driven by scandals. Over the years, multiple anti-corruption technologies, including those using AI, have been developed by integrity techies. This will be discussed in Chapter 3.

Notes

1 The anthropologist and historian Lilia Schwarcz (2019) noted that the phrase "sea of mud" had been used before as a synonym for impunity by the newspaper *Gazeta de Notícias*, which reported that justice in the imperial regime had been "buried" and "everything was a sea of mud." The newspaper referred to the case known as the "robbery of the Crown jewels" which took place in 1882, at the end of the empire, a period during which a Brazilian dynasty of Portuguese origin ruled Brazil from 1822–1889. Crown employees were found to be responsible for the theft but were pardoned and released from jail.

2 In 2020, after being investigated under the auspices of Operation *Lava Jato*, Odebrecht announced that the conglomerate would henceforth be known as Novonor, which is a combination of the Portuguese words *novo* (new) and *norte* (north).

3 Vargas committed to the following: first, the creation of disciplinary commissions to investigate and punish the misuse of public funds; second, the removal of "corrupt agents"; and third, the reorganisation of the judiciary to ensure its independence (Lagunes et al., 2021b). See also http://www.biblioteca.presidencia.gov.br/presidencia/ex-presidentes/getulio-vargas/discursos/discursos-de-posse/discurso-de-posse-1930/view (accessed on August 5, 2022).

4 During President José Sarney's administration (1985–1990), the *Folha de S. Paulo* newspaper revealed allegations of irregularities in the public bidding for the construction of a railroad, leading to a congressional inquiry (*Ferrovia Norte-Sul*) and the cancellation of the bidding process (Freitas, 1987; Gaspar, 2020). The *Anões do Orçamento* scandal emerged at the beginning of the 1990s and was linked to the diversion of public funds to favour legislators and construction firms, among them Odebrecht (Matais et al., 2016; Gaspar, 2020). The scheme involved public money being directed to ghost firms controlled by the members of the congressional budget committee, most of whom were physically short of stature, and members of the so-called *baixo clero* (low clergy), hence the tactless nickname "dwarves" (Praça, 2011).

5 *Lava Jato* was an investigation into money laundering that became public knowledge in March 2014 and quickly turned into a much bigger corruption probe, uncovering a wide and intricate web of political and corporate racketeering with unprecedented repercussions not only in Brazil but in several Latin American and African countries (Lagunes et al., 2021b).

6 The origin of this saying dates back to the 1960s when staff members of the Brazilian football team Palmeiras, founded by Italian immigrants, left a tense 14-hour meeting to go to a pizza restaurant in São Paulo. The journalist Milton Peruzzi, who was following the imbroglio, came up with the following headline in the newspaper *Gazeta Esportiva*: "Palmeiras Crisis Ends in Pizza" (Meneghetti, 2017). Since then, the slang phrase "it ended up in pizza" has been used when something very serious happens, such as a crime or a scandal, but nothing is done about it, and everyone just accepts the non-conclusive outcome.

7 For an in-depth explanation of each agency's role, see Aranha (2020), who mapped out the formal role and interconnection of the Brazilian accountability institutions, delineating how they establish ties to effectively combat corruption and impose administrative and legal penalties on those engaged in activities such as procurement fraud, misappropriation of public funds, and inflated contracts.

8 See http://www.planalto.gov.br/ccivil_03/decreto/1990-1994/D1048impressao.htm.

9 In 2022, Lula, then aged 77, was elected for a third term after serving a prison sentence on corruption charges.

10 See https://www1.folha.uol.com.br/folha/brasil/ult96u612372.shtml (accessed on July 15, 2022).

11 The kick-off for the creation of the electronic voting system happened years before, in 1986, with the consolidation of a unique identifier for registering

voters saved in a digital database aiming to curb fraud. In 1994, the Brazilian electoral court carried out the first electronic processing of the results of a general election. In the following year, the electronic ballot started to be developed with the goal of eliminating fraud in the electoral process by removing the human element from the vote counting process. See https://www.tse.jus.br/comunicacao/noticias/2014/Junho/conheca-a-historia-da-urna-eletronica-brasileira-que-completa-18-anos.

12 For a detailed illustration of the legal apparatus, see OECD (2020, Chapter 2, Figure 2.3).

13 See https://www.gov.br/mcti/pt-br/acompanhe-o-mcti/transformacaodigital/arquivosinteligenciaartificial/ebia-documento_referencia_4-979_2021.pdf (accessed on September 27, 2023).

14 In 2023, the SEI expansion process became responsibility of the *Ministério da Gestão e da Inovação em Serviços Públicos* (Ministry of Management and Public Innovation in Public Services).

3   The rise of integrity techies and their digital technologies

Jorge Jambreiro Filho abandoned his MA in Artificial Intelligence (AI) when he was approved to become a government employee for the *Receita Federal do Brasil* (Brazilian Federal Revenue Service, commonly referred to as the Revenue Service), following the formal entrance examination. During his mandatory training course to become a civil servant in 1997, he had the idea to create an automated AI system to expedite customs selection procedures, a system that in 2014 became the *Sistema de Seleção Aduaneira por Aprendizado de Máquina* (Sisam – Machine Learning Customs Selection System) (Jambreiro Filho 2015a, 2015b, 2019). The system was conceptualised to enhance customs control through the generation of consolidated importer data which assists in post-clearance reviews. With his information technology (IT) and programming skills, Jambreiro Filho developed two tech solutions using the tools available at the time.

In a two-part report entitled *A História do Sisam como a Vivi* (*The History of Sisam as I Lived It*), published on his website,[1] Jambeiro Filho wrote:

> I thought that by using robots, I could obtain the necessary data to accomplish what I had planned from the beginning, without realising that this automation of repetitive and tedious tasks would end up becoming the focus of a controversy spanning more than a decade.

The civil servant shared his journey of encountering steadfast resistance to innovation for more than a decade in public service. He also witnessed projects being halted due to shifts in top-level management (Jambreiro Filho 2015a, 2015b, 2019). The public sector explicitly barred the use of bots and forbade him from creating additional technological solutions. His efforts to innovate resulted in him facing

DOI: 10.4324/9781003326618-3

disciplinary proceedings for the unauthorised use of the Revenue Service database to test his proof of concept, a move that almost jeopardised his job. On the one hand, his story illustrates how closed public administration in Brazil had been regarding civil servants' ideas to develop AI-based technology and introduce it to their governmental bodies' daily routines. On the other hand, it also shows how things have changed in Brazilian public administration over the past few years. In 2023, nine years after its implementation, Sisam was providing inspectors with alerts, visual analytics, and communication tools to combat fraud and tax evasion in import operations. These resources were easily integrated into inspection reports, having been generated automatically using natural language processing (Odilla, 2023a). The Revenue Service also started to craft numerous innovative tech anti-corruption solutions. In addition, in mid-2019, Jambreiro Filho was made head of the Centre of Excellence in Artificial Intelligence of the Brazilian Federal Revenue Service.

The Revenue Service may be one of the first agencies to have a centre to test new ideas and develop innovative solutions in public administration, but it is not the only one. Sano (2020), for example, identified 63 initiatives that call themselves innovation labs operating in the executive, judiciary, and legislative branches, at the federal, state, and local level in Brazil. In addition, Jambreiro Filho is not the only civil servant to have developed his in-house solutions to improve accountability and control within a governmental agency without outsourcing them. Interviews with civil servants from law enforcement agencies in Brazil suggest that numerous digital anti-corruption initiatives were not solely in-house solutions, as previously noted by Odilla (2023), but were also initially crafted by tech-savvy individual civil servants in what one interviewee called a "closed innovation" process.

In a similar vein, numerous bottom-up anti-corruption tools have also been developed by tech-savvy individuals or concerned citizens open to innovation who were helped by volunteers with a background in IT, which facilitated the creation of many digital tools without outsourcing. These bottom-up initiatives have been created not only by individuals but also by civil society organisations (CSOs) and media outlets. They all rely on a substantial volume of public and open data, and increasingly easier access to software and computer programming information. The open public data generated by bureaucratic management has proved to be crucial not only for the country's eventual transition to electronic government (Neves and Silva, 2023) but also for involving citizens in innovative pro-transparency and monitoring initiatives (Odilla, 2023a).

This chapter presents various top-down and bottom-up digital initiatives and shows how they have been applied in anti-corruption measures. While not an all-encompassing compilation of the digital technologies already in place, the initiatives presented here serve as illustrative examples due to their breadth and diversity. The chapter also sheds light on the individuals and organisations responsible for the design, creation, and deployment of these digital technologies. These individuals are referred to as "integrity techies," a diverse group of actors who engage with digital technologies to fight corruption, improve transparency, and reinforce accountability in Brazil. Both digital technologies and their creators must be seen as part of the "web of accountability" (Mainwaring and Welna, 2003; Power and Taylor, 2011), in which, as already mentioned in the previous chapters, several actors, governmental and non-governmental, human and non-human, assume oversight, investigative, and disciplinary roles. Integrity techies are also key to better understand the social dimension of anti-corruption technologies (ACTs), while the digital tools they develop encompass the material dimension.

Top-down anti-corruption initiatives

In Brazil, ACTs developed from the top-down align the formal roles of law enforcement agencies, including the anti-corruption ones, with their internal capacity to develop these digital technologies. Each law enforcement agency at the federal level has its responsibilities defined by legal frameworks, and this directly impacts the functionality of its ACTs. For example, the Federal Revenue Service serves as the tax and customs authority. The *Controladoria Geral da União* (CGU – Office of the Comptroller General), in turn, is responsible for conducting audits and inspecting public funds, imposing administrative sanctions on companies and civil servants, advancing active transparency and the right to information, establishing national networks to enhance public integrity, and encouraging involvement from civil society. The CGU has some overlapping roles with the *Tribunal de Contas da União* (TCU – Federal Court of Accounts) as both are responsible for auditing the accounts and overseeing the implementation of federal budgets, checking cases of non-compliance with public procurement rules and the mismanagement of public funds. Additionally, the TCU scrutinises the accounts of all those who contribute to financial losses, mismanagement, or irregularities affecting the public purse and, as might be expected, the main functions of the digital technologies it uses revolve around its formal duties.

Although these government agencies have developed tools that are aligned with their different roles and are limited to their constraints within the web of accountability, they were often created owing to the vast amount of public data and information available and the increased interest in anti-corruption activities driven by scandals. It is worth stressing that top-down ACTs typically operate in silos, with minimal interaction among agencies taking place regarding their digital solutions. Their codes are not open and are rarely shared with other government units, and the ACTs' inputs and outputs are treated as sensitive data (Odilla, 2023a). Some of the top-down technologies are designed for internal use by civil servants employed by specific agencies, while others are created for people interested in accessing government data or providing information in the form of a compliment, critique, question, or denunciation.

As digital systems are daily becoming increasingly complex, many types of digital technologies can serve multiple purposes, combining various functionalities and technologies. Yet the top-down ACTs can be clustered into similar types. For analytical purposes, the following ACTs comprise social-technical assemblages that are clustered according to their material dimensions, i.e. types of digital technologies: (i) *digital platforms*, used to store, search, and cross-check systematised public data; (ii) *monitoring and risk-detecting bots*, used to assist humans in preventing and identifying wrongdoings, mainly by cross-checking data and raising "red flags"; and (iii) *chatbots*, used to engage with citizens by impersonating humans on major social media platforms. Next, an extended definition of each one of these digital technologies is provided as well as examples of government ACTs that were already in place at the time of writing in 2023.

Digital platforms

Digital platforms can be broadly defined as "a programable digital architecture designed to organise interactions between users, not just end users but also corporate entities and public bodies" (van Dijck et al., 2018, p. 4). By default, they require the digitalisation and standardisation of data. According to van Dijck et al. (2018, p. 9), platforms have automated features, and are organised by algorithms and interfaces. They are also formalised through ownership relations and governed through user agreements. Regarding anti-corruption government digital platforms, they operate as non-profit entities and may be accessible to the public or restricted to civil servants. Their primary purpose is to facilitate various aspects of governance and integrity, including enhancing active transparency and streamlining accountability processes through monitoring, investigative, and disciplinary processes.

One of the first Brazilian anti-corruption platforms was initially designed to be an internal system to manage auditing procedures. *Fiscobras* was developed in the late 1990s by the TCU. Given its status as a long-term law enforcement organisation aiming to improve financial management,[2] the TCU initiated relatively early investments in technology to modernise its operations. It began its digitalisation process in 1977 with the establishment of a data processing centre (TCU, 2014). However, it did not introduce personal microcomputers until the mid-1990s. With these in place, it adopted the Windows operating system leading to the development of the *Fiscobras* platform (*Sistema de Fiscalização de Obras Públicas*, or Public Works Inspection System) for use by the court auditors (TCU, 2014).

Fiscobras stands out as a significant milestone in the realm of digital solutions, not only within the TCU but also at the federal government level. As with many other ACTs in Brazil, its creation was prompted by a scandal and it was established under intense political pressure, mirroring the trajectory of various other digital technologies in Brazil, as discussed in the preceding chapter. According to the TCU (2016), in 1995, a temporary parliamentary committee was convened in the Senate to review any incomplete public projects and works, an issue that has continued to plague Brazil's infrastructure landscape throughout its history. After identifying 2,214 abandoned hospitals, bridges, schools, and other public structures, the Senate recommended that the TCU should audit these projects and initiate ongoing inspections for public construction and engineering endeavours (TCU, 2016). This directive also called for the creation of specialised technical teams or units in this field.

In 1997, to aid Congress in drafting the budget bill and to prevent the misuse of public money by cataloguing construction projects with indications of irregularities already examined by the TCU, an electronic form was created using Microsoft Access. This form facilitated the collection and storage of information amassed during the inspections conducted in the preceding year. The resulting file, stored on a floppy disk dispatched by courier or post to the TCU's state-level technical units along with an audit guide, was intended to be completed with findings, including illustrations. By 1998, a similar form was updated and introduced on the TCU's intranet, to be accessible in real-time to all units, thereby speeding up processes. The platform has undergone progressive refinement over the years thanks to the IT-trained staff who began to be hired in the late 1980s. In 2023, information about incomplete public projects available on *Fiscobras* could be accessed through an interactive open portal (https://paineis.tcu.gov.br/), which includes graphs, tables, maps, and filters. TCU not only has integrated platforms but has also been developing other systems that assist *Fiscobras*, such as the *Sistema de Auditoria de Orçamentos* (Budget Audit System).

In the case of the CGU, it has evolved into a more multifaceted agency that goes beyond auditing public expenditure and public policies. For example, it is also tasked with promoting both active and passive transparency. To this end, in 2004, the CGU launched the *Portal da Transparência*. The open online government transparency portal, developed and hosted by CGU's employees, began to provide timely reporting of budget and expenditure information. In 2024, the *Portal da Transparência* integrated and presented data from various systems used by the federal executive for its financial and administrative management, aiming to provide transparency in public administration and empower society for social monitoring. This data includes, for example, information about the budget, bidding contracts, disbursements for social programmes, civil servants' travel allowances, monthly salaries of government employees, and sanctions imposed on employees in the federal executive branch and on business suppliers (Odilla and Rodriguez-Olivari, 2021).

The portal, notably, enabled the press to uncover irregular governmental expenses involving the use of government payment cards, leading to a political scandal in 2008; since then, it has also been used as a key source to explore cases of misuse of public money. In the case of the *Escândalo dos Cartões Corporativos* (corporate cards scandal) in 2008, the controversies ultimately led to the removal of the cabinet minister who had championed extravagant spending in an airport duty-free shop zone during that period, as well as the resignation of another who used public funds to purchase tapioca pudding. These were unprecedented scenes in Brazil, a country familiar with impunity.

This revelation of the misspending of public money was followed by a Federal Police investigation and a Parliamentary Committee of Inquiry aimed at scrutinising the misappropriation of public funds. Despite the initial turmoil, the portal is still very active. It has gained new databases and functionalities over the years. Not only can data available there be downloaded in the form of comma-separated value (CSV) files but the CGU started to offer the data in an application programming interface (API) format to facilitate automated data collection, cross-checking, and analysis. Many participants working for the CGU interviewed for this research consider the *Portal da Transparência* a milestone in Brazil. It is not only because the *Portal* is a platform that combines various types of data, which required a massive effort in standardisation and management of data and practices, but also because it is open and easily accessible to those with internet access.

The *Portal da Transparência* served as a model for many other government platforms for promoting active transparency across the country, spanning different government levels and branches. Following the corporate cards scandal, new legislation (Complementary Law No. 131) was enacted in 2009, requiring the real-time availability of detailed information on budgetary and financial execution at the federal, state, and municipal levels. The legislation aimed to guarantee fiscal responsibility and to enhance social control of public planning and spending.

The Brazilian Lower House (*Câmara dos Deputados*), for example, has been incorporating open and proactive transparency features on its website, which was initially launched in 1997.[3] In 2004, its web portal implemented a link for "transparency," and since then it has slowly started to release data not only about internal procedures but also its expenditures. Although compensation funds to cover expenditures, such as travel, meals, postage, and stationery for elected representatives have existed since 2001, detailed data have only been available online since April 2009, and previous receipts remain undisclosed. The decision to open them in that year was also made following press reports of the improper use of public funds (Folha de S.Paulo, 2009).

Over the years, the Lower House has improved the accessibility of the data available on its platform and now provides a wide range of detailed, machine-readable information on roll call votes, legislative propositions, party affiliations, parliamentary fronts, caucus composition, public speeches delivered in committees and on the floor, attendance, and expenditure. The Lower House's Open Data platform (*Dados Abertos*, see https://dadosabertos.camara.leg.br/) not only offers data in various formats such as JSON and XML for scraping but provides tutorials and a list of 13 civil society initiatives that have used Lower House data.

Unlike the TCU and the CGU, which have responsibilities related to combating corruption, the Lower House has not made any direct efforts to establish ACTs. However, it appears to serve as an essential data provider to citizens interested in curbing corruption and enhancing the accountability of elected congressional members. This is evident from the perspective of one particular interviewee, a civil servant who is part of LabHacker, a unit of the Lower House created in 2014 to promote the collaborative development of innovative citizenship projects related to the federal legislature The interviewee recognised the limitations of combating corruption from within and emphasised the importance of external control:

> This [model] of exposing [misuse of public funds by elected representatives], of showing society what is happening, is the kind of thing that the House wouldn't do on its own, by itself. It's unlikely that there would be a service within the House that would carry out this analysis. It's politically unviable for this to be approved internally. But it's something that other institutions and civil society initiatives are capable of doing. I think a model has been established very clearly that works for transparency and combating corruption, which is to show what is happening to society as a whole.[4]

There are other remarkable platforms, most of which were not intended to combat corruption per se but have been used as one of the main sources of anti-corruption digital tools. One example is the aforementioned e-procurement platform *ComprasNet*, launched in 1997, which later became the *Portal de Compras do Governo Federal* (Purchase Portal of the Federal Government). It is an open digital archive that provides detailed and updated information about active and potential suppliers, materials, and services, as well as the publication of notices, bidding documents, bidding results, and contract extracts (Otranto Alves et al., 2008). The platform also includes electronic procurement programmes for conducting electronic bidding, electronic price quotations, and support for in-person bidding. In addition, it provides access to statistics and a database of legislation, regulations, and publications related to procurement.

Another case is the *e-proc*, which began as a pilot project in southern federal courts in 2003 and later was expanded to the entire judiciary. Programmed in PHP and using MySQL, the *e-proc* was an evolution of two previous data storage systems developed in the 1990s: the Massachusetts General Hospital Utility Multi-Programming System (SIPRO/MUMPS)[5] and the Sistema de Acompanhamento Processual (Siapro, a system for following up court procedures). The *e-proc* transformed the judiciary by digitalising all sorts of procedures and allowing uploads not only of documents but also wiretaps and court hearings, including the cross-examination of witnesses and the interrogation of defendants. The *e-proc* was developed by IT civil servants, ensuring the complete security of information and at a low cost to the public treasury. The *e-proc*, used to follow court procedures, has proved to be an invaluable source of raw data for those closely following the *Lava Jato* investigation, because not only does it provide online access to seized materials, official reports, and court documents, but also it allows law enforcement and judicial actions to be monitored, mainly by journalists. The *Lava Jato* investigation also benefited from direct access to a wide range of digital databases, including property and vehicle registries, and the various technological advances promoted by the National Network of Anti-Money Laundering Technology Laboratories established in 2007.

The CGU, the federal-level Brazilian anti-corruption agency, has also been investing in other types of digital platforms. *Fala.BR* (see https://falabr.cgu.gov.br/), for example, combines services for requesting access to information, reporting wrongdoing involving the public administration, requesting a service delivery, or making a compliment, a complaint, or a suggestion. It is coordinated by the *Ouvidoria* (Ombudsman) and offers tutorials, and access to a temporal series of data, such as requests to access information in alternative formats (e.g. CSV, XML). The *Dados Abertos* portal, in turn, can be seen as a data lake, i.e. a centralised repository of standardised data, mainly government data, in the form of a platform. It has been built up and improved in several phases since 2011 by civil society representatives at tech events and festivals such as the Campus Party.[6] There are also limited access platforms, albeit that these are less common. The CGU created, for example, the *e-PAD*, which has been specifically designed for use by internal affairs units. It seeks to enhance the management of administrative disciplinary procedures for civil servants (the CGU-PAD) and accountability procedures for companies (the CGU-PJ). These platforms serve as essential tools to investigate and discipline individuals and companies engaging in corruption and other wrongdoings.

Monitoring and risk-detecting bots

Bots are "obedient agents following their developers' programming," with "a central infrastructural part of computer architecture and the internet" (Monaco and Woolley, 2022, pp. 2–3). In other words, they are software programs, most of which are run on computer hardware, that are designed to perform the monotonous tasks that humans do not enjoy or cannot do quickly, helping to maintain services, gather and organise vast amounts of online information, perform analytics, and send reminders (Monaco and Woolley, 2022). By 2023, most of the top-down bots in place in Brazil were already AI-based anti-corruption tools, defined as data processing systems driven by tasks or problems designed to, with a degree of autonomy, identify, predict, summarise, and/or communicate actions related to the misuse of position, information and/or resources aimed at private gain at the expense of the collective good.

(Odilla, 2023a)

The Revenue Service was one of the first government agencies to start developing solutions primarily to support human activities, addressing the constraints of limited human and financial resources for audits, inspections, and other law enforcement tasks (Odilla, 2023a). One example of this monitoring and risk awareness tool is the *ContÁgil*, an in-house solution developed by a civil servant from the Revenue Service, that, since its launch in 2009, has been incorporating new data retrieving and processing tools and functionalities based on the most common supervised learning algorithms (decision trees, naïve Bayes, support vector machines and deep neural networks) (Jambreiro Filho, 2019). Its clustering, outlier detection, and topic discovery functionalities are mainly related to identifying fiscal fraud and money laundering. *ContÁgil* can perform in the space of an hour what a human inspector would take a week to do and can cross multiple internal and external datasets and build network graphs indicating the level of relationships between people and companies (Jambreiro Filho, 2019).

Analisador Inteligente e Integrado de Transações Aduaneiras (*Aniita* – Intelligent and Integrated Analyser of Customs Transactions) and *Batimento Automatizado de Documentos na Importação* (*BatDoc* – Document Mismatch Detector) are also innovative technologies that stand out as internal solutions created by the Revenue Service. Since its inception in 2012, *Aniita*, for example, has provided integrated and intelligent software for seamless searches, as documented by Coutinho (2012). It leverages various datasets related to customs transactions, including sensitive and protected data such as import declarations and digital images of auxiliary documents (such as invoices and bills of lading). Its primary purpose is to identify anomalies associated with potential cases of customs duty evasion and avoidance. In addition, it employs optical character recognition to process digital images of auxiliary documents, recognising relevant fields and normalising data to accommodate variations in the way these fields are presented in different documents. Developed in Java and utilising Abby, a commercial optical character recognition tool, *Aniita* features a user-friendly dashboard that highlights disparities in key aspects such as company names, addresses, and prices. (Odilla, 2023a). This comparison is made between the goods' descriptions in the invoices and their corresponding entries in the import declaration. *Aniita* has now been integrated into the Sisam system, which was introduced earlier in this chapter.

The CGU also witnessed the emergence of different types of digital tools developed by civil servants, designed as internal solutions mainly to speed up their daily tasks. One such tool was created to identify company ownership. It gained popularity among other government employees both within and beyond the creator's unit. The system underwent updates, incorporating dozens of data sources, and eventually evolved into *Macros*: an online searchable system that centralizes crucial information for inspections and audits. It was launched in 2012.[7] This application offers the ability to download spreadsheets and access readily available reports, complete with graphs and tables. *Macros* also includes a network analysis tool for establishing connections between individuals and companies. This innovation was recognised through an award in the 20th innovation contest organised by the *Escola Brasileira de Administração Pública e de Empresas* (ENAP – Brazilian National School of Public Administration).

When recounting the story of *Macros*, its creator, who holds an engineering degree and joined the CGU for the civil servant status without having a precise understanding of the agency's responsibilities, explains how everything started in 2006 with a pile of paper.

> A demand was made ... to perform a cross-reference, and it was all on paper, initially. The cross-reference had to do with the fight against corruption ... They handed me a gigantic table full of paper that I couldn't even see across. ... Doing this by hand would take a year, it was impossible to do this thing manually. So, I proposed to them: let me study it here, how the system works, what the system does, and how can I automate it, since I came from the control and automation field. I already had experience. It was the first task I did here, and then I never left the office because they wouldn't let me go.[8]

The interviewee explained that he had to learn how to use Visual Basic to program and, later, he enhanced the solution to transmit the data collected from the website of the Revenue Service to an Excel spreadsheet. Subsequently, he established network connections using Graphviz to generate, at that time, static images. "It was old school," he remarked. However, other civil servants from different units began approaching him to perform searches for owners and business partners. He had a waiting list for consultations and his program was informally

spread throughout law enforcement agencies. The civil servant emphasised the pros and cons of giving civil servants the autonomy to seek
technological solutions to their daily challenges:

> I had already switched units and gone to Goiás. I was already
> working on something else there and one day a guy from the
> Federal Police in Rio Grande do Sul called me, asking: "Did you
> create this script?" I said I did. He replied: "I'm using it here." It
> has been about ten years since any maintenance was done, the
> script was returning errors, but he wanted to know how to use it.
> So, the script evolved and started to be used by other agencies –
> the Public Prosecution Service used it, the police used it, everyone
> used it. It was cool. … The thing is that our solutions here, they all
> end up being like this (in-house). I had my own work issues, and I
> had the tools [to solve them]. … I didn't hand it over to someone
> else to create and test it. I started doing it and adapting as I went
> along. Then someone else would bring up a good idea, and it
> would fit right in. It is a much more organic thing. You know, in a
> normal software development process, you give it to someone to
> make it. That's the advantage; let's say, the process flows much
> faster. The disadvantage is that it's not sustainable. When things
> scale up, you lose control. Someone out there is running into
> issues, and then it starts to blow up and affects everyone.[9]

To enable civil servants to access various databases, the interviewee
obtained an older computer to serve as the server, utilised Linux, an open-
source Unix-like operating system, and collaborated with other colleagues
to create the *Macros* system which evolved into an interactive database.
The civil servant kept developing other tools, such as the e-Audit, to
manage the auditing procedures. The system integrates, within a single
electronic platform, the entire audit process, from the planning of control
actions to the monitoring of issued recommendations and the recording
of benefits. More recently, the creator of *Macros* developed a database
that has, as an output, a dashboard with a wide range of information
regarding the education sector.

Since 2015, both the CGU and the TCU have invested in algorithms
designed not only to cross-check data but also to alert users about
potential risks within a specific context. The primary aim of the *risk
awareness bots* is to deliver timely information and insights regarding
risks, enabling users to make well-informed decisions and to take
appropriate actions. Utilising machine learning techniques, the bots
conduct risk analyses, categorise potential risks, assess their severity,

and estimate their impact. When a significant risk is detected, the bot promptly issues alerts and notifications to relevant stakeholders, which can take the form of emails or colourful tags on dashboards. These include in-house tech solutions such as the well-known bots Alice, Monica, Sofia, Carina, Agata, and Adele. With the exception of Alice, which is shared by the TCU and the CGU, the others are exclusive to the TCU.

Most of these names are acronyms linked to the primary tasks of the ACTs. For example, Alice stands for *Analisador de Licitações, Contratos e Editais* (Analysis of Bidding and Calls for Bids), which sends emails to auditors before public bids are made when it detects any suspicious findings. Monica stands for *Monitoramento Integrado para o Controle de Aquisições* (Integrated Monitoring for Procurement Control). It offers data visualisation tools with filters and the option to download spreadsheets containing information about monitoring public procurement activities in the federal judicial and legislative branches. Sofia stands for *Sistema de Orientação sobre Fatos e Indícios para o Auditor* (Guidance System on Facts and Evidence for the Auditor). Sofia reviews draft texts by verifying sources and references, identifying correlations between the information written in the text and other procedures, and providing findings in comment boxes to support civil servants when writing audit reports.

Carina stands for *Crawler e Analisador de Registros da Imprensa Nacional* (Crawler and Analyser of the National Gazette), developed to identify anomalies related to urgent public health-related contracts and bids. It offers a dashboard and email alerts with information about suspicious purchases. Agata stands for *Aplicação Geradora de Análise Textual com Aprendizado* (App for Generating Textual Analysis with Learning). It was designed to refine the list of words in public procurement documents and their respective textual contexts that may require further inspection. This is achieved by utilising human input to train the algorithm and incorporate the results into other monitoring tools. Adele stands for *Análise de Disputa em Licitações Eletrônicas* (Dispute Analysis in Electronic Bidding) and highlights inconsistencies and anomalies in electronic public bidding through a dynamic dashboard format (for a detailed explanation of the respective data inputs, processing and outputs of all these bots, see Odilla, 2023a).[10]

In 2023, the TCU launched ChatTCU, a generative AI tool inspired by ChatGPT that summarises case documents directly from the TCU's online platform, allowing auditors to ask questions and receive answers. The tool was outsourced to a private consortium, one of the few times that the TCU has opened a public tender for technological innovation. ChatTCU is hosted on a dedicated instance of Microsoft Azure OpenAI Service's cloud platform. It has been used for document analysis, legal research, translation, and administrative consultations. Similar tools have been developed by other Court of Accounts at the regional level. ChatTCU also sparked interest from the Contraloría General de República de Chile (Chilean Office of the Comptroller General) and the Inter-American Development Bank, that initiated talks to adopt similar solutions using the TCU's code.

In general, there is a very low level of collaboration among public agencies to share tools, codes, and data (Odilla, 2023a). The biggest exception is the bot Alice, which was created in 2015 by the CGU to assist auditors in combating corruption in public procurement, embezzlement, graft, and anti-competitive practices by analysing bid submissions, contracts, and calls for bids. The tool was seldom used by the CGU until the TCU expressed an interest in its code. One year after the inception of Alice, an agreement was signed to share the code with the TCU, which enhanced the tool by incorporating machine learning techniques (a random forest classifier) alongside the existing regular expressions (regex) for daily data mining from the Federal Official Gazette and the *ComprasNet* portal (Odilla, 2023a). Alice's codes are now being shared with regional goverments.

Top-down chatbots

Top-down anti-corruption chatbots are less popular compared to other types of government bots. Designed to interact with citizens, they are found on social media platforms. Zello, a chatbot developed by the TCU was initially active on Twitter but later migrated to WhatsApp and was also accessible through the TCU mobile application on Android and Apple iOS systems. Zello was developed to communicate via text messages and to assist citizens in accessing information about the TCU, including details about auditing procedures and individual clearance certificates. The name of the bot is derived from the word *zelo*, and is intended to convey a sense of care and concern for others' wellbeing.

The CGU has also ventured into developing chatbots for engaging with people on social media. One such bot, named Cida (*Chatbot Interativo de Atendimento Cidadão*, or Interactive Chatbot for Citizen Service), was implemented in Java and initially used Facebook Messenger and Telegram to interact with users. Cida's primary function was to guide users in submitting complaints, reports, greetings, and feedback through the web on the *Fala.BR* platform, which serves as an ombudsman and crowdsourcing platform, and is also used for requesting access to information. Cida was discontinued due to limitations imposed by Facebook for chatbots and relatively low usage on Telegram. These two cases show that relying on mainstream platforms may have disadvantages as it can make government chatbots susceptible to external technological constraints as well as fluctuations in platform popularity.

One interviewee,[11] a civil servant who works as an auditor but is also an activist involved with local monitoring groups, noted that many useful solutions utilising emerging technologies still have limited access. He argued that the most advanced tools should be made available to the public or for use by anti-corruption CSOs in the form of chatbots. He regrets that the Brazilian federal executive has been focusing more recently on granting access to digital data rather than providing almost ready-to-use monitoring and risk awareness digital tools. Some civil servants interviewed for this book, including the participant mentioned above, believe that the state needs to do more to create better digital solutions for citizens or to help them directly to develop or use digital technologies, mainly to monitor local government activities. According to the interviewees, there are weaker accountability mechanisms in municipalities. This is not to say that civil society is entirely separate from the process of the digitalisation of anti-corruption that has facilitated the development of anti-corruption digital solutions in government agencies. On the contrary, as will be demonstrated in the section that follows, bottom-up digital technologies have also been created.

Bottom-up anti-corruption initiatives

The increasing access to digital data and technological devices has inspired not only civil servants but also tech-savvy and concerned citizens who have embarked on using their tech skills to develop digital technologies from the grassroots. Although bottom-up ACTs that deploy digital technology to fight corruption in Brazil are fewer in number and often have a shorter life when compared to top-down ones, they are also innovative and present diverse functionalities. These types of initiatives depend heavily on public data that has been made available by the government through digital platforms or obtained through access to information requests (Odilla, 2023a). As expected, bottom-up ACTs are consistent with the expected roles of civil society in the web of accountability theoretical framework. In Brazil, they are primarily designed to constrain political power by increasing citizens' knowledge, awareness, participation and mobilisation beyond elections (Smulovitz and Peruzzotti, 2000).

Again, the initiatives detailed next are clustered according to the material dimension of ACTs. In the case of bottom-up initiatives, there are four types of digital technologies with some overlaps and multiple uses: (1) *desktop and mobile monitoring applications*, used primarily to monitor the performance of public officials and promote transparency; (2) *"integrity guard" bots* that deploy emerging technologies, such as AI-based tools, to automate the collection and analysis of the use of public money or performance of public agencies or sectors; (3) *chatbots*, used to engage and mobilise users willing to hold public officials or governmental actions to account; and (4) *existing social media platforms*, used to expose suspicious misuse of public money and other wrongdoings as well as to mobilise and organise civic actions.

Desktop and mobile monitoring applications

Bottom-up anti-corruption applications started to be designed in the early-2000s. They first emerged in the form of websites. Over the years, they have gained additional functionalities, such as filters, maps, or ranks. More recently, this type of technology has been offered in the form of mobile applications. They have been created to cover various aspects of public life that can be considered relevant for those monitoring public officials, such as attendance at legislature voting sessions, travel expenditures, and bills proposed and approved in the Lower House and Senate. Other ACTs list the potential involvement of politicians in corruption, sometimes compiling reported corruption cases in news outlets, and at other times collecting data directly from court websites. For example, *Políticos do Brasil, Às Claras*, and *Perfil Político* were developed in the form of a website to offer access to consolidated public open information on candidates running for elections including, for example, campaign donations and candidates' assets declarations, based on digital data available on the Electoral Court and Congress websites. *Parlametria* and *Meu Deputado* were also created to gather information about the performance of elected politicians, but they started out as mobile apps. *Corruptômetro, Excelências*, and *Deu no Jornal* provide information about scandals published by legacy media or court procedures.[12] Some of these apps are currently in use while others have been discontinued.

The non-profit organisation *Transparência Brasil*, initially established in the early 2000s as a chapter of Transparency International in Brazil, deserves recognition as a pioneer in the development of bottom-up anti-corruption technologies in the country. *Excelências* (2001–2017), *Deu no Jornal* (2004–2012), and *Às Claras* (2006–2021) were some of the first bottom-up ACTs. All of them use different sorts of public open data available at the time they were created and aimed at increasing transparency and improving accountability. Not only was *Transparência Brasil* the first Brazilian CSO to develop ACTs but it has continued to invest in new types of technology over the years. Its trajectory suggests a co-evolutionary path between digital media and grassroots anti-corruption efforts as the organisation has consistently been led by digitally literate activists and individuals open to digital innovation with a medium to high level of tech capabilities (Odilla, 2024). The participants pointed out that the connection between *Transparência Brasil* and ACTs can largely be credited to one of its early directors, the journalist Claudio Weber Abramo. Abramo had a profound understanding of technology. With a degree in mathematics, he chose a career in journalism and is recognised as the pioneer of data journalism in Brazil. However, the separation of *Transparência Brasil* and Transparency International also occurred during Abramo's leadership. Following this, *Transparência Brasil* began to seek grants to develop new products that were linked to the newly emerging technological tools. The organisation has always employed tech-savvy individuals and people open to experimenting with new technologies.

Bottom-up (social media) chatbots

Transparência Brasil developed, for example, the chatbot Edu, to engage citizens to promote accountability in the education sector, thanks to the tech skills of a data scientist with a background in law who was already working there as a team member. He created a WhatsApp bot to invite people to review school meals and monitor the construction of schools and nurseries in Brazil, as part of the initiatives *Tá de Pé Merenda* and *Tá de Pé*, respectively. The bot was designed to provide friendly, concise instructions and it allowed users to share their location for selecting schools or worksites that were near them for conducting inspections by taking pictures and reporting any issues that were identified. The decision to make a chatbot on the hyphenation seems odd. was taken as an attempt to "be where the user is," according to one interviewee.[13]

Tá de Pé was initially named *Cadê Minha Escola?* (Where Is My School?) and was sponsored by Google, which recommended a company to help *Transparência Brasil* to design the prototype. Working full-time for an intensive five days with specialists from different areas, including two Google employees (both from the marketing team, one specialising in brand management and another in technology) they changed the name to *Tá de Pé* and developed the mobile app that garnered a lower-than-expected number of downloads and interactions. Because of this, Google suggested the transition to WhatsApp, as reported by the participants interviewed for this study.

The chatbot Edu, an abbreviation for education, worked with numbered and clickable options, allowing users to interact with the bot and select, for example, the type of monitoring used during the construction of schools and nurseries or the safety of schools after the ending of the lockdown restrictions imposed due to the outbreak of the COVID-19 pandemic. It allowed users to share their location or select a state where they wanted to conduct monitoring. In a friendly and interactive manner, Edu offered a tutorial and guided users as if there had been a previous conversation, a strategy the creator of the chatbot learned to use for getting messages approved by WhatsApp. Edu was designed to utilise the same back-end system, which was the existing technology infrastructure already in operation for the app. The back-end system, however, was outsourced first to a company recommended by Google and later to an individual who saw the initiative on TV and offered to help. It not only automated the collection of open data regarding the official school construction schedule but also created digital solutions for managing and processing data sent by users participating in the monitoring process. Additionally, the back-end system semi-automated the process of contacting authorities responsible for school construction in the event of user-reported delays. Unfortunately, according to those who participated in the technology creation and implementation, most authorities simply ignored the emails sent asking for clarification.

"Integrity guard" bots

With the greater availability of open data and easier access to technology, civil society actors have begun to develop more advanced tools, including AI-based technologies. This includes bots to monitor and maintain integrity by detecting and flagging suspicious or illicit activities, such as the misuse of public funds by congressional representatives, inspecting the purchase of school meals, and monitoring delays in the Supreme Court.

Tá de Pé, for example, was expanded to monitor the purchase of food for schools and the quality of the meals through a tool named *Tá de Pé Merenda*. It was developed to use AI and data analysis techniques to monitor local food purchases for schools and to provide a website equipped with a search engine for anyone interested in identifying signs of irregularities. AI assists in evaluating factors such as higher-than-average prices, the consideration of seasonal variations in food prices, and the involvement of local producers. The initiative's website highlights a range of unusual purchases and contracts that merit closer examination, but only for the states of Rio Grande do Sul and Pernambuco. Furthermore, the initiative had planned to launch a WhatsApp chatbot to engage with students and encourage them to monitor the quality of school meals by reporting issues and capturing images of meals. Although initially slated for a 2020 launch, the COVID-19 pandemic interrupted the launch plans. And since mid-2023 Edu, the chatbot, has become unresponsive on WhatsApp.

The first Brazilian bottom-up bot using AI was created in 2016 by three tech-savvy friends who, it is worth noting, had never been party to any anti-corruption movement nor defined themselves as activists at the time the initiative was created. But they wanted to use their IT skills and knowledge to do something after Brazil witnessed a significant corruption scandal and the impeachment of President Dilma Rousseff. *Operação Serenata de Amor* (OSA – Operation Love Serenade) was designed to track expenditures made by elected members of the Lower House and to identify the suspicious use of public funding based on high costs and irregular spending. OSA's creators developed a bot named Rosie programmed in Python to scrape and cross-check open data released by the Lower House and other types of data available online, such as the restaurant review site Foursquare, and to tweet automatically when suspicious expenditures were identified, mainly related to irregular payments and overpriced meals. The Twitter posts link them to a dashboard named *Jarbas*, also created by the initiative, to allow everyone with access to the internet to check their receipts. For around six years, the initiative sustained a community of over 600 computer and data scientists, journalists and IT developers who interacted on a Telegram channel, later transferred to Discord. It attracted 40,000 followers on Twitter and 66,000 on Facebook (Savaget et al., 2019; Odilla and Mattoni, 2023). On GitHub, Serenata's open project counts 102 contributors.

A few politicians had interacted with the bot on Twitter. Some of them denied irregularities, others admitted irregularities and promised to pay the misused money back. Despite its innovative features and having succeeded in recovering some public money, since November 2022, Rosie has stopped tweeting. Even fully automated initiatives like this face sustainability issues. In the case of Rosie, its creators struggled to secure funding to sustain the project, which was initiated via a crowdfunding campaign. They left the project but, instead of killing Rosie, they opted to transfer the bot to the non-governmental organisation (NGO) Open Knowledge Brazil (OKBR), which does not treat it as a priority (Odilla and Mattoni, 2023). In January 2024, a post explained that OSA had evolved into other projects, and although the database collected by Rosie would remain available online, OKBR's focus was now on other initiatives.

Over time, Brazil has witnessed the proliferation of numerous Twitter bots, primarily due to its previously friendly and generous API policies for such technologies.[14] One notable example is the Rui Barbot, established in 2018 by JOTA, a non-mainstream news media outlet that also offers paid data analysis services aimed at predicting significant decisions, primarily within the legislative and executive branches. This Python-based crawler was developed by a tech-savvy, with a law degree, to aggregate public data from selected procedures posted on the Supreme Court website. It monitors the last update of each procedure, automatically tweeting about anniversaries if a procedure remains inactive for at least 180 days or even for years. Additionally, it was created to notify JOTA journalists about procedure anniversaries via email. However, its last tweet dates back to June 2020, prompting questions about the sustainability of such initiatives and the heavy reliance on mainstream social media platforms. The bot's name draws inspiration from the words of Brazilian judge Rui Barbosa, who once remarked, "delayed justice is not justice but rather qualified and manifest injustice."

Even initiatives that prioritise human action, such as *Operação Política Supervisionada* (OPS – Operation Supervised Politics), have used integrity guard bots. Since 2013, OPS has relied on volunteers to conduct civic audits of congressional representatives' expenditures. With the assistance of tech-savvy volunteers, OPS has developed digital tools to cross-check and provide comprehensive information, including rankings and automated emails about politicians incurring the highest expenses, to support their monitoring work. The group counts on around 200 volunteers and does not have any physical headquarters. Essentially, OPS exists in a Telegram group. The initiative has also been experimenting with new tools to engage ordinary citizens. Its ranking of congressional expenditures is an open-source project accessible through an interactive website (ops.net.br/), primarily programmed in C# (C-Sharp), and available on GitHub. OPS has already launched an app to allow people to classify the expenditures of elected representatives as either suspicious or unsuspicious. The initiative has also built a Twitter bot[15] to reach a broader audience by collecting data from the Lower House website and displaying images of any congressional figures found to incur high expenditures. At March 2024, the most recent post published by the bot on Twitter was dated June 12, 2023. In addition, a customised prototype of ChatGPT, trained with data on elected representatives' spending, was created but not released for use.

Existing social media platforms

Social media platforms have largely been used by many bottom-up ACTs, even those that have developed their tools. The existing platforms serve different purposes, depending on their affordances and popularity. Twitter, as already mentioned, was the preferred choice due to its friendly and free API service that allowed automation. It was also considered in Brazil to be a social network with a relevant opinion-forming role. Facebook, for years, was the most popular platform and, therefore, it was used by initiatives to promote their actions. Facebook also saw a proliferation of self-named anti-corruption groups and had a prevailing role in creating awareness and mobilising people: it was found that 73 groups created between 2011 and 2020 had the word corruption in their names. Social media, particularly Facebook, also played a relevant role in the emergence of the so-called new right groups, for example, the *Movimento Contra Corrupção* (MCC – Movement Against Corruption), *Movimento Brasil Livre* (Free Brazil Movement), *Vem pra Rua* (Come to the Streets), which organised massive anti-corruption demonstrations in 2015 and 2016.

In the case of OSA, OPS, and *Transparência Brasil*, Facebook, and later Instagram, were used mainly to give visibility and to promote their tools and findings. OPS and OSA used Telegram as their "headquarters." Most of their interactions have been online. The creator of OPS also uses YouTube to promote the outcomes of its monitoring task force. OPS started with a seller showing in a tutorial on YouTube how to use public data to check congressional members' expenditures and to identify suspicious cases (Odilla and Veloso, 2024).

Social media is seen by activists and developers as a cheaper way to reach wider audiences, as exemplified by the well-known campaign to pass the *Ficha Limpa* (Clean Slate) bill, led by an electoral judge with the support of the Catholic Church, the national bar association and a few associations related to law enforcement. In 2009 and 2010, the *Movimento de Combate à Corrupção Eleitoral* (MCCE – Movement to Combat Electoral Corruption) adopted, what were considered at the time, innovative online tactics, ranging from overloading politicians' email boxes to uploading hashtags and broadcasting live events on social media (mainly Orkut, Facebook and Twitter), in order to gather signatures and pressurize congressional members to endorse the bill. This legislation approved mandates that candidates are barred from seeking public office for eight years if their convictions for a designated list of crimes, including corruption charges, are decided by a committee or upheld on appeal (Mattoni and Odilla, 2021, p. 1135). Through existing social media, bottom-up initiatives have effectively raised awareness, garnered support, earned official recognition, and successfully organised large-scale street protests, particularly in 2015 and 2016 (Odilla, 2024).

It is worth noting that the successful online strategy adopted by the MCCE to approve the anti-corruption bill led it to take a step further from just using existing social media. Together with the *Instituto de Tecnologia e Sociedade* (Technology and Society Institute), it developed the free mobile application *Mudamos*, which enabled Brazil's citizens to participate in lawmaking by proposing their bills and signing onto one another's proposals using verified electronic signatures. The app, launched in 2017, was based on blockchain technology to offer secure and verifiable digital signature, and the initiative also included support for proposing, analysing, and improving proposed bills, as well as in-person events such as the *Virada Legislativa* (a legal draft-a-thon) with legislatures, mainly at the municipal level. *Mudamos* was discontinued in 2023 More recently, civil society initiatives have also used existing technologies to collectively create open data lakes or repositories of public data, collected, cleaned and organised in different formats, ready to be downloaded, not only to facilitate access to public data, but also to ensure that once public data is made available, for whatever reason, it remains accessible and preserved for future use. Examples of these repositories are the. *Base dos Dados*, Brasil.IO, *Minha Receita*. (see Chapter 5 and the online Appendix[16] for more details).

Integrity techies

When analysing the process of digitalisation of anti-corruption in Brazil, it is possible to observe the emergence of what are called "integrity techies," that is, actors who have developed or facilitated the development and use of digital technologies. In some cases, this was an individual decision; in others, it was stimulated by CSOs or law enforcement agencies and their newly created innovation labs or dedicated units to use data mining and other emerging technologies such as AI. For both the top-down and bottom-up initiatives studied in this research, integrity techies use technology to directly or indirectly fight corruption, for example by increasing accountability and transparency. These individuals and their digital technologies interact and, more importantly, the latter cannot exist without the former. Integrity techies are shaped by and also shape and influence various aspects of public integrity through their technologies.

Integrity techies are not exclusive to Brazil. This concept is introduced here to define technologically inclined individuals working for non-profit organisations or governments, mainly law enforcement and anti-corruption agencies, who drive the digitalisation of anti-corruption efforts in any given place. Individual motivations are expected to be polysemic, not mutually exclusive. Institutions are important for their rise. But the individual agency is crucial. Individual incentives range from "doing something good" with their knowledge to finding solutions to technological challenges, improving tech skills, curriculum vitaes, and GitHub profiles, speeding up procedures with automated solutions, holding public officials and governments accountable, promoting political empowerment through the digital world, changing the future of the country.

Integrity techies seize opportunities to use their knowledge and technical skills to automate tasks and create digital solutions to improve or facilitate integrity. Both developers and facilitators need to be open to technological innovation. They are therefore involved in the development of a wide range of entities such as bots, chatbots, applications, and platforms to detect, report, crowdsource, expose, and predict instances of corruption. These actors are also expected to be committed to upholding ethical standards in the public sphere through the development and use of technology. This requires individual agency as well as resources, including technical, financial, and government support, as most of them work with public data.

Integrity techies do not operate in a vacuum. They are part of the political landscape, influenced by political discourse and policy decisions, which are often shaped by perceptions of corruption and corruption scandals as well as by the expectations related to digital innovation. On the political front, the role of integrity techies can extend to active engagement in policy frameworks related to access to information and open data, promoting active transparency and influencing decision-making processes. This includes advocating for new laws and regulations that prioritise ethical behaviour and encourage digital innovation. While the practices of integrity techies may differ, their goals are related to strengthening principles for the effective functioning of the accountability web through the development and use of digital technology.

Moreover, their role and approach to anti-corruption largely depend on the web of accountability framework and the legal apparatus already in place, with their digital technologies varying accordingly. For example, depending on the accountability apparatus in any given country, integrity techies working for anti-corruption agencies may focus more closely on sanction enforcement than on oversight activities due to their formal responsibilities, or vice versa. Integrity techies from CSOs are more likely to carry out monitoring activities than investigative activities due to limited access to data, resources, and capacity to conduct investigations. Although this is not the case in Brazil, some CSOs or tech-savvy individuals do conduct investigations. A pertinent example is Bellingcat,[17] an independent collective of researchers, investigators, and citizen journalists in more than 20 countries, most of them volunteers. They use open-source intelligence techniques to investigate a wide range of issues of public interest and have helped to solve major crimes, including the downing of Malaysia Airlines Flight 17 over Ukraine (Higgins, 2021).

Integrity techies in Brazil

The proliferation of integrity techies in Brazil is directly linked to the digitalisation of public administration at the federal level and easy access to programming information. In Brazil, integrity techies are often tech-savvy individuals, many of whom have a background in engineering, computer science or data analysis, or they are technology enthusiasts who have learned computer programming or who want to support technological advances in the fight against corruption. Most state actors involved in initiating or facilitating the use of technology in anti-corruption are career civil servants, and some of them have created tools based on their academic studies, including masters degrees and/or PhDs. This means that, as career civil servants, they have passed very competitive formal exams requiring candidates to hold at least a bachelor's degree, earn high salaries, and enjoy job stability and special employment rights (Odilla, 2020a). It is worth noting that both the TCU and the CGU had specific formal examinations for candidates with a background in technology and computer systems and this also attracted tech-savvy people to these law enforcement agencies.

While those working in governmental initiatives earn high salaries and enjoy job stability, integrity techies engaged in civil society initiatives aim to make a living from civic action but struggle to do so, and often do not stay long on the civic tech projects. In addition, they are more actively contributing to the digitalisation of the fight against corruption in the public administration and the three branches of power at the federal level, although there are initiatives focused on municipalities (as in the case of *Querido Diário*, which, thanks to the work of volunteers, collects, processes, stores, and creates search tools for data from local government gazettes). It should be noted that the digital technologies developed by the integrity techies could also be used in the private sector to curb corporate corruption, although this is not the focus of this book.

As mentioned, both integrity techies and their technologies are closely aligned with their formal responsibilities in the web of accountability framework. For example, those working for the CGU have invested more in monitoring and investigative technologies, and tools to allow and assess citizens' interactions with the agency. This is because the agency combines both preventive and law enforcement functions, including close work to promote citizen participation (Odilla and Rodriguez-Olivari, 2021, 20–21). While the CGU has developed digital tools for both internal and external users, the Revenue Service has developed sophisticated technologies that benefit its employees' daily routines, specifically in relation to the monitoring of government revenue derived from the movement of goods in customs and from the collection of taxes. Within civil society, the course of action of integrity techies often hinges on the availability of funding and public open data. However, their efforts extend to disseminating information about politicians and public administration with the intent of empowering voters to hold incumbents accountable by withholding their votes during elections.

In Brazil, integrity techies also engage among themselves and in activities to share knowledge and offer training to interested citizens, researchers, journalists, and other professionals in the tech field, fostering a culture of collaboration and continuous learning within the technology community. Organisations such as Open Knowledge Brazil, responsible for OSA, *Transparência Brasil*, and even the CGU offer training on how to access and use public data and conduct civic monitoring. Open Knowledge also provides training on the use of Python for civic innovation, for example.

If digital data and access to technology are pre-conditions, being part of an environment where innovation is promoted can be seen as a facilitator. This has been happening in Brazil. Civil society has seen a proliferation of hackathons being organised by different branches of government at the federal and state level to create incentives for tech savvies to create digital solutions to fight corruption. OSA has also organised coding sprints in the form of face-to-face and virtual events to speed up the solution of technical issues and to report their initial suspicions to the Lower House.

In public administration, there have also been efforts to encourage civil servants to innovate and showcase their initiatives. One example is ENAP, which recognizes innovative technological tools through awards and hosts an innovation week in public administration. The TCU not only arranges workshops and meetings, inviting civil servants, civil society, companies, and even academics, but also operates the Institute Serzedello Corrêa which provides training and higher education courses for its civil servants, fostering innovation and the generation of solutions to enhance daily agency operations.

Additionally, agencies such as the TCU and the CGU have also witnessed a substantial number of employees pursuing Masters and PhD degrees, both in Brazil and abroad, as mentioned earlier. Associated with tech skills, there is the essential resource for any digital tool: digital data. One interviewee[18] summarised this as follows: "It's like giving children an amusement park. ... It has become a big video game, but for finding solutions, generating useful information for decision-making, and preventing, combating, and detecting corruption."

From the data collected, I observed a non-technocentric view among those creating digital technologies. Participants highlighted how they perceived tech-human relations as complementary while emphasising that technology alone is not sufficient to combat corruption. However, for most of them, technology serves as an enabler of innovation and a tool to empower citizens. This is directly associated with their goals. The data indicates that among those most desired are: increasing citizen engagement in anti-corruption activities; including indirect activities such as improving transparency and enhancing accountability; and being willing to serve society with their expertise. It was also observed that digital technologies developed are human-centric. Most bots and chatbots have personal names and are represented by stylised (almost childish) representations of a robot, or they communicate through friendly and interactive outputs such as dashboards and emails.

In Brazil, integrity techies' main motivation in engaging with ACTs has less to do with the anti-corruption fight and more to do with seeing the creation of technologies to benefit their careers or simply to use their knowledge and skills as a way to innovate. Among the governmental employees, suppressing staffing shortages and speeding up procedures also served as a primary motivation. This phenomenon occurred, for example, with IT developers and those seeking job stability in the civil service. Many integrity techies, however, have started to view themselves as activists after creating or deploying their ACTs. Some civil servants have set up NGOs or assisted in forming collective actions, and some IT developers who joined initiatives as volunteers have established their pro-transparency initiatives or have begun to advocate for access to data, for example. As a result, some of them have become activists, although not necessarily anti-corruption activists, but rather data or open data activists.

Conclusion

This chapter discussed the digitalisation of anti-corruption efforts in Brazil, presenting a wide range of top-down and bottom-up initiatives and highlighting the emergence of "integrity techies," actors who influence and are influenced by social and political aspects of integrity. Key actors in understanding the digitalisation of anti-corruption, integrity techies can be viewed as tech-savvy individuals, often with a background in engineering or computer science, or as people who are more open to technological innovation who make efforts to create digital solutions for anti-corruption objectives in individual projects, collective actions, CSOs, or government agencies. These solutions, such as monitoring and risk-detecting bots, chatbots, monitoring or open data platforms and applications, are seen by those creating them as complementary but not sufficient to fight corruption, reflecting the human-centred approach of integrity techies in the development of digital anti-corruption tools to date. For these actors, access to technology and digital data is crucial. It was also observed that top-down initiatives represent a form of closed innovation, as they are more likely to be related to the work of law enforcement agencies and often deal with sensitive data. Bottom-up initiatives tend to have broader objectives and rely more on public open data. Both top-down and bottom-up initiatives are frequently developed in-house, with a very low level of outsourcing, at least in the Brazilian case.

It is noteworthy that Brazilian legislation requires governments to offer information in a machine-readable format, aiding the digitalisation of the anti-corruption fight, at least at the federal level, as the public administration makes efforts to digitise its data and citizens can easily scrape them. The state, however, has not only been a data supplier. In the next chapter, an analytical tool that emerged from empirical data is introduced to evaluate various forms of interaction between state actors and civil society in the battle against corruption. In addition, I delve into the factors contributing to the proliferation of integrity techies in the country. Interestingly, in Brazil, civil society anti-corruption organisations do not provide platforms for reporting and exposing corruption, unlike practices in other countries where platforms such as "I Paid a Bribe" in India (Chakraborty, 2024) serve this purpose. This topic is also explored in the chapter that follows.

Notes

1 See https://www.jambeiro.com.br/jorgefilho/ (accessed on August 16, 2023).
2 The TCU was created with the name of *Tribunal de Contas* in 1890, becoming fully operational three years later. Its status as one of the most important control agencies at the federal level was subsequently reiterated in all of Brazil's constitutions – in 1891, 1934, 1937, 1946, 1967, and 1988 – and over time it has undergone gradual improvements. Since 1988, it has been tasked with assisting Congress to account for and oversee the implementation of the federal budget and governmental performance (Brito, 2009). It is worth noting that, although created in 1890, the TCU has its roots in the imperial regime (1822–1889). Initially, internal control of revenue expenditure and financing was supposed to be enforced by the Royal Public Treasury, created by a royal order from King Dom João VI, and later by the National Treasury, a court created in 1924 and restructured in 1931 (Odilla, 2020b).
3 For the evolution of the Lower House web portal, see https://www2.cama ra.leg.br/sobre-o-portal/historico (accessed on October 9, 2023).

4 Interview conducted online on December 15, 2020.

5 MUMPS was a database initially created for managing data in the health system but from 1978 it started offering a commercial version for companies.

6 For more information about the creation of the *Dados Abertos* portal, see https://www.gov.br/governodigital/pt-br/dados-abertos/portalbrasileirodadosa bertos.pdf (accessed on October 11, 2023).

7 For a detailed explanation of *Macros*, see https://repositorio.enap.gov.br/ bitstream/1/2567/1/Sistema%20Macros.pdf (accessed on October 11, 2023).

8 Interview conducted online on January 4, 2023.

9 Interview conducted online on January 4, 2023.

10 The predominance of female bot names is noteworthy. A civil servant involved in Agata's development and Alice's upgrade initially joked that they are an "everlasting tribute to women." However, a more detailed explanation relates to the units creating the tools. Alice, for example, originated at the CGU, and while there was a discussion about changing its name, it was ultimately retained to maintain consistency. This naming trend emerged, with acronyms and names of women guiding product choices. Some find certain names forced. Notably, different departments have varied naming preferences. For example, the Secretary of Technology and Information tends to favour corporate names, while the *Secretaria Geral de Controle Externo* (Segecex –General Secretary for External Oversight) opts for more open names. Interview conducted online on February 2, 2022.

11 Interview conducted online on January 9, 2023.

12 *Corruptômetro* is an app that offers a collection of news on corruption published in the mainstream media. See https://apps.apple.com/in/app/corruptom etro-brasil/id584205560. *Perfil Político* allows users to make comparisons of candidates or to obtain information about specific individuals based on data regarding their individual trajectories, asset declarations, and gender and race representation. See https://perfilpolitico.serenata.ai/. *Excelências* offered an online database with information initially on congressional members and later members of local assemblies, but it was discontinued. *Parlamentria* offers different databases to monitor the Congress. See https://parlametria.org.br/proje tos. The app *Meu Deputado* allows the performance of different members of the Lower House to be compared. See https://meudeputado.mobi/. *Deu no Jornal* was a compilation of corruption-related news articles from mainstream media outlets, but it is not in use anymore. *Às Claras* aggregated information about electoral financing for each candidate since 2002, but it was discontinued. All weblinks were accessed on August 22, 2023.

13 Interview conducted online on December 3, 2021.

14 After Elon Musk acquired Twitter (now rebranded as X) in 2022, one of his initial significant actions was to change the company's expansive free API service for third-party developers and researchers and to start to charge for its use.

15 See on X (formerly Twitter) *OPS Fiscalize* @NFiscais, launched in June 2022.

16 See F. Odilla (2024). *Anti-Corruption Technologies in Brazil: Online Appendix*. Available at https://osf.io/7gzwu/.

17 See https://www.bellingcat.com/about/who-we-are/ (accessed on March 1, 2024).

18 Interview conducted online on December 22, 2022.

4 The material, social, symbolic, and political dimensions of anti-corruption technologies in Brazil

In 2023, Brazil was already being recognised as a "major innovator" in employing emerging digital technologies designed to optimise the allocation of audit resources and proactively prevent corruption (Murray et al., 2023). Indeed, the country has witnessed the emergence of a wide range of anti-corruption technologies (ACTs). Both governmental and civil society "integrity techies" have been actively involved in developing and implementing novel digital technologies, as illustrated in the previous chapter. The integrity techies leverage extensive open government data to bolster transparency, reinforce accountability, and combat corruption. Additionally, most of the digital tools already in place were not outsourced; rather, they were developed in-house by tech-savvy civil servants and concerned citizens. As mentioned earlier in this volume, the digitalisation of anti-corruption in Brazil was made possible by previous digitisation and digitalisation processes that resulted in a wide range of digital databases enabling access to machine-readable public data.

Corruption scandals have significantly influenced not only the political rhetoric (Lagunes et al., 2021a) but also important reforms, including those geared at the digitalisation (and digitisation) of public administration in response to specific scandals, as shown in Chapter 2. In Brazil, many reforms tend to come from within the bureaucracy (Rich, 2019, 2020), sometimes with the support or partnership of civil society, as was the case with the two most relevant legal changes in the anti-corruption realm (Praça and Taylor, 2014; Da Ros and Taylor, 2022). As noted by Morris (2021), most Brazilian anti-corruption reforms focus on the gradual strengthening of the capacity and autonomy of law enforcement agencies, in particular those with anti-corruption roles. These features render the executive and legislative powers less important and the process more gradual, sequenced, and piecemeal (Praça and Taylor, 2014; Da Ros and Taylor, 2022; Morris, 2021, p. 84).

DOI: 10.4324/9781003326618-4

Can the same be said about the widespread development of ACTs in Brazil? Are executive and legislative powers less important than bureaucratic commitment in the case of ACTs? If so, why has the country witnessed a rapid digitalisation of anti-corruption through a proliferation of innovative integrity-driven technologies? The aim here is to answer these questions by assessing the key dimensions of ACTs in Brazil. This chapter takes as its starting point the framework proposed by Mattoni (2024) for grassroots ACTs, extends it to top-down initiatives, and adds the political dimension to the material, social and symbolic dimensions that support these socio-technical assemblages in anti-corruption practices.

The chapter first offers a more analytical perspective, starting with a reflection on the main differences between top-down and bottom-up initiatives. It builds on the detailed description provided in the previous chapter and applies a "situated and pragmatic" analytical approach (Mattoni, 2021) that helps to illustrate the importance of the "web of accountability" framework, embedded in both the polity and politics and explains how integrity techies and their digital technologies position themselves in the anti-corruption field. Next, the chapter assesses each of the three main dimensions of the Brazilian ACTs as proposed by Mattoni (2024) and adds a fourth one, the political dimension. As socio-technical-political assemblages, ACTs are a compilation of various symbolic, material, social, and political elements that come together when digital technologies are incorporated into individuals' or collectives' anti-corruption practices. This analysis is conducted to provide elements that, when seen together, can help to explain the emergence of Brazil as a prominent innovator harnessing new digital technologies to assess corruption risks within the public sector. The chapter concludes with some reflections on the importance of six crucial elements that not only have allowed integrity techies to emerge but have also ensured the rapid diffusion of ACTs among law enforcement agencies and civil society organisations (CSOs) engaged in anti-corruption practices.

Bottom-up versus top-down anti-corruption initiatives: a situated and pragmatic approach

Having a situated and pragmatic outlook on anti-corruption initiatives facilitates an understanding of embedded contextualities as well as their practical, real-world influence (Mattoni, 2021). Here, a "situated" approach means taking into consideration what corruption is in particular contexts, as well as the environment in which digital technology is being developed and used for fighting specific types of corruption. Like corruption, there are multiple situations in which technology is imagined, developed, and then employed. They are also tied to various corruption and anti-corruption scenarios (Mattoni, 2021). If a "pragmatic" outlook is adopted it emphasises a practical, action-oriented approach that prioritises accessible and usable data and meaningful and actionable results, such as raising awareness, speeding up procedures, and saving time and effort, stimulating and promoting upward and downward transparency, horizontal and bottom-up accountability, and governmental integrity.

The situated-pragmatic lenses proved useful for analysing Brazilian ACTs. From a top-down perspective, as mentioned in the previous chapter, law enforcement agencies have been, as expected, developing digital technologies with functionalities and affordances closely related to their formal attribution and responsibilities. This means that, for example, the *Tribunal de Contas da União* (TCU – Federal Court of Accounts) focuses more on technologies designed to identify the risk of misuse of public funds, particularly in public procurement, as there are more accessible and long-established databases, while the *Receita Federal do Brasil* (Brazilian Federal Revenue Service, commonly referred to as the Revenue Service) prioritises the development of digital technologies for clearance procedures in customs departments and cases of tax evasion and money laundering. As the Controladoria Geral da União (CGU – Office of the Comptroller General) conducts public audits, enters lenience agreements, and investigates federal civil servants, it has been developing a wider range of digital solutions, from "red flags" in public contracts and risks of corruption among civil servants, to an open-source platform for registering and monitoring public meetings. In addition, these agencies often do not share their technologies, even in the form of codes, or their data inputs or outputs, claiming that legislation does not allow them to do so.

In the case of the CGU, as noted in Chapter 3, there is also noticeable concern regarding both "upward transparency" and "downward transparency." The former refers to creating channels and technologies for citizens to openly share information, feedback, and concerns about the government, while the latter typically involves the flow of information or communication from the government to citizens, ensuring that decisions, processes, and goals are shared openly and effectively (Adam and Fazekas, 2021). Agencies such as the TCU and the Revenue Service, in turn, possess more closed initiatives exclusively designed for internal use. Although both offer channels for requesting information and submitting complaints or compliments, these channels are less developed in comparison to the numerous internal systems and bots already in place to assist inspectors with their duties. One exception is the TCU's chatbot Zello, which provides information on procedures and issues clearance documents in the Court of Accounts.

As noted by Odilla (2023a), while most government ACTs have access to both sensitive and open public data, bottom-up initiatives often rely on limited access to open data. It is not a coincidence that the scope of ACTs from the grassroots is narrower, leading to a more extensive use of social media when compared with top-down ACTs. Many initiatives fight corruption indirectly by improving transparency, raising awareness, and mobilising people to participate in monitoring actions, protests, or policymaking. Still, as is the case with government ACTs, most bottom-up ACTs aim to tackle the misuse of public money. Sometimes this is done directly, such as the ACTs that monitor politicians' expenditures, including *Operação Serenata de Amor* (OSA – Operation Love Serenade) and *Operação Política Supervisionada* (OPS – Operation Supervised Politics), or the application to check the construction of schools that are behind schedule, as the non-profit organisation *Transparência Brasil*'s *Tá de Pé* was designed to do. Others tackle corruption indirectly by raising awareness of criminal and administrative procedures in which politicians are defendants, or exposing politicians and parties on social media. More recently, initiatives have focused on the transparency of public data. As mentioned in the previous chapter, initiatives are creating centralised repositories to store structured public data.

However, there are some notable exceptions regarding non-governmental initiatives that focus on other areas not related to the misuse of public money. One of them was the Rui Barbot, established in 2018 by JOTA, which identified pending cases at the Supreme Court that have been waiting to be heard for months and even years. In the case of ACTs focused on the judiciary, it is worth noting two ACTs that have been not mentioned thus far: *Dados Jus Br*, an initiative launched in 2020 by the Federal University of Campina Grande and *Transparência Brasil*, which organises and collates data on salaries and top-up salary payments in the bodies that make up the Brazilian justice system and also offers a searchable dashboard with graphs and tables. Another example is a tool named *Publique-se*, created by the chapter of Transparency International in Brazil and Abraji (Brazilian Association of Investigative Journalism) and outsourced to a company called Digesto to collect information on legal proceedings that cite politicians as defendants or plaintiffs in the higher, federal, and local courts.

The key dimensions of ACTS in Brazil

The decisions made by both bottom-up and top-down ACTs regarding the focus on particular types of actions and the targeting of specific forms of corruption, extend beyond mere considerations of data availability or personal motivation. They are linked to at least four dimensions, which together explain the different types of ACTs. Mattoni (2024) listed three of them: (1) A *material dimension* related to digital technologies, including available and assessable data, software and hardware, employed to counter corruption and their affordances; (2) a *social dimension* related to interactions among a wide range of actors at both the individual and organisational level that impact the type of relationship between the state and civil society; and (3) a *symbolic dimension* related to imaginaries of how anti-corruption efforts should be performed. This book adds a fourth *political dimension* related to the existing institutional arrangements in political systems that influence not only whether but also the way digital technologies are employed to fight corruption from the top-down and the bottom-up. In the anti-corruption domain, these arrangements manifest themselves in the "web of accountability" framework, and are intricately integrated into both the polity and politics, indicating how integrity techies and their digital tools position themselves and establish their formal responsibilities.

That is why ACTs cannot be seen as synonymous with digital technologies applied in anti-corruption efforts. Complex in nature, ACTs can be conceptualised as socio-technical-political assemblages with, as stressed by Mattoni (2024), the overarching and more abstract goal of addressing corruption issues through digital technology. ACTs' more immediate and concrete aim is to target specific types of corruption and related behaviours, as discussed in the previous section. ACTs can be seen, therefore, as continual undertakings that materialise from interactions involving diverse elements, incorporating both human and non-human actors, and constantly developing over time (Mattoni, 2024). Therefore, ACTs extend beyond the technological realm and are not neutral. Next, these four dimensions are assessed taking into consideration the Brazilian case.

The material dimension

The material dimension, as noted earlier, is related to digital technologies, including their data inputs, processing and outputs, as well as the hardware involved and their affordances. Some anti-corruption efforts engage with the development of highly sophisticated digital technologies, such as artificial intelligence (AI)-based ones, that are exclusively designed to curb corruption (dedicated devices and infrastructures). Others rely on existing technologies, such as social media platforms, that have been created for other purposes but are also used in anti-corruption (non-dedicated devices and infrastructures). Yet others are embedded in complex ecosystems, mixing novel and current technologies, dedicated and non-dedicated devices and insfrastructures, such as OSA/Rosie (Odilla and Mattoni, 2023). For example, digital technologies can be designed to be deployed on desktop and mobile devices, using more or less popular computer languages to process data and relying on a vast amount of both public and private data. While bottom-up ACTs depend on open data, top-down ones also use sensitive and protected data. It is evident that ACTs encompass new and existing technologies, and, depending on the technological elements used, they can be associated with different ideal types of data practices (Mattoni, 2017, 2021), as data serves as the essential raw material of any digital technology.

Data creation involves generating data on corruption that does not yet exist, often through dedicated crowdsourced applications such as platforms that collect reports on wrongdoings. It also includes compiling new datasets that are already publicly available or exposing classified data. *Tá de Pé* and *Faro* are examples of dedicated applications that create new data. The former effectively transforms citizens into civic inspectors who gather information on school construction by taking pictures and collecting data. The latter produces new data for the government by analysing complaints and reports submitted by citizens to determine which issues require further action and which should be shelved.

OSA and OPS, on the other hand, have dedicated infrastructures to compiling new datasets based on public and private open data to identify suspicious expenditures by congressional representatives. The dedicated bots that cross-check different datasets and create risk scores for the TCU and the CGU also fit into the typology of data creation. This is the case of the monitoring and risk detection bots such as Alice, Monica, Sofia, Carina, and Adele, and risk systems such as the *Sistema de Seleção Aduaneira por Aprendizado de Máquina* (Sisam). These tools developed for monitoring and detecting risks or actual cases of corruption often offer their outputs in user-friendly formats, such as X (formerly Twitter) posts, dashboards, or email alerts combining current and newly created information. Therefore, not only do they involve data creation but also *data transformation*.

Data transformation relates to the reorganisation and conversion of data into information that is more accessible to a wider audience (Mattoni, 2021). Numerous dedicated ACTs developed by the Revenue Service exemplify this practice. For instance, Sisam uses natural language processing to generate comprehensive reports based on data from inspected and non-inspected import declarations. This allows for the calculation of around 30 potential errors within each import declaration. Another example is the monitoring and risk awareness tool *ContÁgil*, which leverages multiple internal and external datasets to construct network graphs that depict the extent of relationships among individuals and companies. This tool accomplishes tasks that would take a human inspector a week to complete in the space of an hour (Jambreiro Filho, 2019). The bot Alice provides its outputs in the form of alert emails and a dashboard with interactive maps, graphs, and tables. The CGU has also provided interactive panels and charts of data through the open online government transparency portal *Portal da Transparência* for the general public and through the *Macros* for its auditors. OPS similarly engages in data transformation by ranking members of the Senate and the Lower House (*Câmara dos Deputados*) according to their expenditures. These rankings include photographs of politicians, their party affiliation, state, and expenditures. Additionally, the initiative offers charts that compare the annual values allocated for covering congressional expenses.

Finally, *data curation* pertains to the incorporation of data, mainly through non-dedicated platforms, into performative practices. In this context, data can act as a catalyst for public outrage and other collective actions. For instance, when *Transparência Brasil* publishes its periodic reports based on data analysis of pressing topics such as judiciary payrolls, financial misconduct in the education and health sectors, and parliamentary budget amendments, they often attract media attention and gain visibility. This dissemination of data serves a performative function. Apart from exposing suspicious cases, it can be designed to engage and mobilise users who are willing to hold public officials or governmental actions to account. OSA used the data collected and analysed by Rosie on Twitter to encourage the bot's followers to scrutinise any members of parliament suspected of misusing public funds. Looking from a top-down perspective, Zello was a chatbot that also employed embedded data when sharing clearance certificates or information on procedures targeting individuals via instant messaging apps or direct messages on Twitter or Telegram.

Table 4.1 outlines various data practices and technological tools used in both top-down and bottom-up approaches to improve

Table 4.1 Material elements of ACTs in Brazil, based on ideal types of data practices and their overall scope, by top-down and bottom-up initiatives

ACTs (who developed them)	Data practice Overall scope	Primary function in the accountability cycle (oversight, investigation, corrective measures)	Material Tech devices and infrastructure	
			Dedicated	*Non-dedicated*
Top-down	*Data transformation* Store systematised public data and allow rapid search to cross-check information for suspicious connections and wrongdoings	*Oversight, investigation*	Digital platforms (e.g. *Fiscobras, Portal da Transparência, ContÁgil, Macros* etc.)	
	Data creation Automate the detection of irregularities and suspicious actions aiming at preventing and identifying wrongdoings	*Oversight, investigation, corrective measures*	Monitoring and risk-detecting bots, mainly using AI-based applications (e.g. Alice, Monica, Sofia, Carina, Adele, Sisam, etc.)	
	Data creation Engage with citizens by impersonating humans to collect reports and denounces and/or share information on corruption cases	*Oversight*	Chatbots embedded in governmental portals	Chatbots on social media platforms (e.g. Zello and Cida)

ACTs *(who developed them)*	Data practice *Overall scope*	Primary function in the accountability cycle *(oversight, investigation, corrective measures)*	Material *Tech devices and infrastructure* Dedicated	Non-dedicated
Bottom-up	*Data curation* Monitor the performance of public officials and promote transparency	Oversight, investigation	Desktop and mobile oversight applications (e.g. *Politicos do Brasil, Às Claras, Perfil Político, Corruptômetro, Parlametria,* and *Meu Deputado*)	
	Data creation Automate the collection and analysis of the use of public money or performance of public agencies or sectors aiming at raising red flags when detecting suspicious wrongdoings	*Oversight, investigation, and pressure for corrective measures*	"Integrity guard" bots, with increasing AI-based applications (*OSA, OPS, Tá de Pé*)	
	Data curation Expose suspicious cases, engage and mobilise users willing to hold public officials or governmental actions to account	*Oversight, pressure for corrective measures*		Chatbots on social media platforms (*OSA/Rosie, OPS/ RobOPS, JOTA/Rui Barbot, Transparência Brasil/Tá de Pé*)
	Data curation Organise mobilisation and sustain participation	*Oversight and pressure for corrective measures*		Instant messaging apps (*OSA* and *OPS*'s *Telegram group, OKBR's Discord*), video

ACTs (who developed them)	Data practice / Overall scope	Primary function in the accountability cycle (oversight, investigation, corrective measures)	Material / Tech devices and infrastructure — Dedicated	Non-dedicated
	Data curation / Organise mobilisation and sustain participation	Oversight, investigation, and pressure for corrective measures	Tweets by Rosie/OSA and WhatsApp interaction by *Tá de Pé* to ask people to oversee public expenditures, e-petition platforms (MCCE with Avaaz in the *Ficha Limpa* campaign)[1]	Social media platforms (all bottom-up initiatives mapped by this present study)
	Data curation / Expose suspicious misuse of public money and other wrongdoings and organise civic actions	Oversight, investigation, corrective measures		
	Data transformation / Store systematised public data to allow rapid search and crosschecking	Oversight, investigation	Data lakes or data repositories collectively collected, cleaned and constructed (e.g. *Base dos Dados, Brasil. IO, Minha Receita*)	

Source: Table created by the author, based on Mattoni (2024, 2021, 2017).

transparency and accountability in public affairs, detailing their overall scope and primary function in the accountability cycle (oversight, investigation, corrective measures) as well as the types of devices or platforms employed.

As illustrated in Table 4.1, most of the ACTs were created using dedicated infrastructures and, as already noted, most of them were developed by tech-savvy citizens and government employees, mainly due to the financial constraints of outsourcing this work. The ACTs also benefited from large amounts of governmental digital data, often available in machine-readable formats, although there are still limitations, as will be discussed in the next chapter. It can also be observed that, in many cases, the same ACT can be designed to engage in different types of data practices, while also making use of both dedicated and non-dedicated infrastructures in complex tech ecosystems. In addition, as expected, data curation is a practice that is more common in bottom-up initiatives. However, what has been seen in Brazil is the direct participation of law enforcement agents, such as auditors, inspectors, and controllers, not only in creating their own ACTs but also in joining forces with civil society actors to promote performative practices such as attempts to change legislation, granting access to data, and offering guidance related to public expenditure monitoring. In Brazil, the social interactions that sustain the creation and use of ACTs in anti-corruption practices involve fewer contentious or coercive features, as opposed to a high level of dependency on the state, as explored next.

The social dimension

Overall, in Brazil, state bureaucrats have played a positive role in supporting independent forms of civic organisation and mobilisation (Rich, 2019), especially since the country's transition to democracy in the 1980s. In many cases, it served as a means for civil servants to cultivate allies and exert pressure for change, bypassing potential obstacles arising from a lack of political will among elected officials (Rich, 2019, 2020). In the case of anti-corruption, the findings suggest that many activists have cultivated fewer confrontational relationships with anti-corruption and law enforcement agencies, particularly the prosecution service, the CGU, and the TCU. The reverse is also true. Over the years, Brazil has witnessed these agencies strengthening their ties with CSOs, although during President Jair Bolsonaro's administration (2019–2022) the number of partnerships between governmental agencies and CSOs, including anti-corruption organisations, dwindled.

State actors have provided training, offering seats on committees and working groups to discuss open data, transparency, and integrity policies, and organising conferences and meetings to facilitate exchanges (Odilla, 2024).

In the 2000s, for instance, the *Adote um município* (Adopt a Municipality) initiative was established. Under this programme, anti-corruption inspectors volunteered to act as "godparents," serving as advisers who assisted concerned citizens in organising efforts to audit and monitor public procurement at the local level. This initiative received initial support from Auditar (the Association of Auditors of the Federal Court of Accounts). In 2005, Auditar and other like-minded organisations founded the *Instituto de Fiscalização e Controle* (IFC – Institute for Oversight and Control), a non-governmental organisation (NGO) created to monitor and control public expenditures. The IFC collaborated with other non-governmental and governmental agencies to launch the *Caravana Todos Contra a Corrupção* (Caravans against Corruption) initiative. As explained by Odilla (2024), in this initiative, groups consisting of civil servants and civil society members with expertise in accounting and integrity travelled to various municipalities to engage with local authorities, organised civil society representatives, and ordinary citizens, discussing topics such as corruption and transparency. These efforts aimed at, and to a large extent succeeded in, mobilising civil society to enhance social and societal controls at the municipal level. These initiatives counted with the help of *Amarribo* (Associated Friends from Ribeirão Bonito), one of the first groups that started fighting corruption from the grassroots in Brazil at the municipal level. Created in 1999, *Amarribo* became known nationally by organising workshops and distributing more than 120,000 booklets that helped to establish almost 100 similar institutions throughout the country (Odilla, 2024).

Over the years, with both support and pressure from civil society, bureaucrats have also actively sponsored events to strengthen civil society participation not only in monitoring but also in designing and advocating for new anti-corruption legislation. Bureaucrats also largely benefit from activism and civil society's support to help them to pursue their own agenda and policy goals. The case of the creation of the *Lei de Acesso à Informação* (LAI – Access to Information Law) is telling in this regard. The idea came from *Transparência Brasil*, which formed a coalition with other CSOs for advocacy. But the bill was also embraced by the CGU, which helped to draft and approve the bill, implement the new legislation, and guarantee its enforcement. Since 2011, when the LAI was approved, the CGU has been authorised to rule on appeals against decisions made by any federal agencies that deny access to information and has been offering an online database with all requests and replies.[2]

Another example is the *Conferência Nacional sobre Transparência e Controle Social, Consocial* (National Conference on Transparency and Social Control), which brought together around 1,300 representatives from various grassroots anti-corruption organisations held in Brasília in 2012. This event, organised and sponsored by the CGU, revolved around discussions regarding enhancement proposals for anti-corruption mechanisms. Additionally, in the same year, the 15th International Anti-Corruption Conference was collaboratively arranged by Transparency International and other Brazilian NGOs with a resolute stance against corruption and an emphasis on integrity. Hosted in Brasília by the government of Brazil and the CGU, the conference attracted more than 1,900 participants representing the public, private, and non-profit sectors hailing from 140 nations (Transparency International, 2012).

It could be observed that this enduring connection with law enforcement agents and civil society actors has also influenced the strategies of well-established anti-corruption civil society initiatives and those striving to remain non-partisan and free from political polarisation. Some of the events promoted by the CGU also resulted in new partnerships and ACTs, as noted by one interviewee,[3] who mentioned the case of *Transparência Brasil* and the Lab Analytics from the Universidade Federal de Campina Grande (Federal University of Campina Grande) which together developed the *Tá de Pé Compras* to monitor public procurement.

In general, CSOs and grassroots movements are inclined to prioritise dialogue with state actors rather than resorting to a confrontational approach as their primary strategy. The cases of OSA and OPS are telling in this regard. When examining why both initiatives adopted a "name and shame" strategy, Odilla and Veloso (2024) observed that exposing politicians and their suspicious misuse of public funds on social media was not the activists' initial approach. This strategy was only pursued after both initiatives failed to persuade Congress and law enforcement agencies to investigate the initiatives' monitoring findings. The activists interviewed revealed that they initially presented cases of public fund misappropriation to the authorities; however, their response was surprisingly disheartening. Instead of opening procedures, the authorities encouraged activists to go to the media and make their findings public. Hence, according to the activists, the "name and shame" approach was less about the intention to damage the reputations of public officials and more about an effort to compel authorities to take sanction-related actions. In many cases, the strategy worked. Both initiatives not only managed to gain visibility and attract new supporters after being portrayed as corruption fighters in the mainstream media, but in a few instances they also succeeded in prompting politicians to reimburse money or even to undergo investigation and eventual conviction.

Yet, despite the anecdotal evidence and our findings, we still lack sufficient theory on public resources and the opportunities given by state bureaucrats to help CSOs to participate in their anti-corruption and pro-transparency agendas. This is particularly true of interactions in which digital technologies play an important role. Table 4.2 is, therefore, an attempt to fill this gap. It brings ideal types of interactions considering the roles of both human (civil society and state) and non-human actors. Digital technologies are incorporated in the table because the material dimension shapes and is shaped by the social dimensions of ACTs.

As illustrated in Table 4.2, the various ideal types of interactions can be situated along a spectrum, with partnership and sponsorship representing the favourable end, and repression and co-optation at the less desirable extreme. This distribution considers the context of interactions and power dynamics, paying attention to the role of digital technologies. The roles of each human and non-human actor, however, vary not only in purpose but also in their level of influence, contribution, and interaction within the broader context of the system or scenario under consideration.

The findings suggest that state actors played crucial roles in the anti-corruption fight from the grassroots in Brazil. It is worth saying that the role of anti-corruption state actors is more prominent in sponsoring and combining efforts with bottom-up initiatives than with incumbents in the executive and legislature. As technological access advances, however, the role of state actors seems to be changing from state-promoted activism to state-open data suppliers, mainly for those initiatives that have been using digital media to oversee governments and public officials.

In addition, in the case of Brazil, state co-optation and repression are less likely to be observed. One exception must be mentioned: justice Dias Toffoli, one of the 11 justices on the Supreme Court, unilaterally ordered an investigation into Transparency International (TI) in Brazil for alleged mismanagement of public funds. This decision has raised serious concerns about the risks of lawfare for NGOs fighting corruption in the country. This is not to say that anti-corruption initiatives have not been used for (anti)partisan purposes – and it has happened in Brazil, for instance, with the emergence of groups that make considerable use of social media and develop more contentious repertoires against specific political parties, their administration at the national level, their partisans and political allies, as though they were the only ones responsible for all the corruption and malaise in the Brazilian political system (Montevechi, 2021). This is the case of the so-called new

Table 4.2 Types of interactions considering the roles of state actors, civil society, and digital technologies in the fight against corruption

Type of interactions (based on the level of confrontation, from none to high)	Roles			Examples of initiatives in Brazil, if available
	State actors	Civil society	Digital technology	
Partnership	Foster coalitions and co-design solutions to improve accountability mechanisms	Engage in coalitions, often with law enforcement agents, to bolster oversight, enhance transparency, and aid open government. Be prepared to use confrontational tactics, if needed	While its use is often combined with more analogue activities, such as training sessions and meetings, tech can be employed to accelerate the process of data production and transformation	Adote um Município/ IFC-Auditar and Amarribo; Observatório Social do Brasil; Transparência Brasil
Sponsorship	Offer resources and opportunities to help civil society to achieve its goals without any kind of co-production	Enjoy subsides but manage to keep being independent and using pressure tactics to achieve their goals	Although its use is often combined with more analogic tactics, tech can be used to accelerate the process of data production and transformation	Amarribo; Observatório Social do Brasil
Data/Information Supplier	Provide reliable data via transparency portals, and freedom of information laws without any kind of co-production	Benefits of already available public data to pursue its anti-corruption goals	Allows data transformation and replication on other media (data embedment)	Rosie/Operação Serenata de Amor; OPS - Operação Política Supervisionada; Rui, the bot; Publique-se; Dados Jus Br
Neutral	Adopt a hands-off approach	Organised interests flourish or flounder on their own, avoiding tensions or confrontations	For the production of data and attracting visibility	Transparência Brasil

Type of interactions (based on the level of confrontation, from none to high)	Roles			Examples of initiatives in Brazil, if available
Cleavage	Divide, divert, or compete with social movement organisations	Constantly challenge governments and/or state actors	Used mainly to transform data, gain support and attract visibility	New "right" collective actions *Vem para Rua, Movimento Brasil Livre, Movimento de Combate à Corrupção*
Responsive	React positively to the pressure, taking action	Advocate for feasible demands and organise sensible and responsible actions. May use contentious tactics if necessary	Used mainly to produce and transform data to attract public authorities' attention	*Movimento de Combate à Corrupção Eleitoral* and the citizens' popular petition that resulted in the *Ficha Limpa* law
Repressive	Oppress (with physical violence or not) civil society initiatives by imposing costs and causing moral outrage within a broader population	Energised permanent resistance that undermines authorities and exposes acts of misconduct in more contentious ways	Used to transform and embed data aiming to confront power elites	
Co-optation	Offer subsidies and other benefits to convert civil society into docile and obedient allies	Stop using public pressure tactics to achieve their goals and become submissive	Used mainly to replicate official content	

Source: Table created by the author.

Note: The empty cells show that non-relevant examples were identified among the cases mentioned in this study.

right that emerged in the mid-2010s, among them groups such as *Vem pra Rua* (Come to the Streets), *Movimento Brasil Livre* (Free Brazil Movement), and the *Movimento de Combate à Corrupção Eleitoral* (MCCE – Movement Against Electoral Corruption). It is worth noting that, although organising demonstrations against the administration of the Partido dos Trabalhadores (Workers' Party), these groups kept closely aligned with bureaucrats of the prosecution service and the judiciary directly involved in the *Lava Jato* (Car Wash) investigation.

Apart from these anti-corruption groups that employed disruptive strategies, such as street demonstrations, and made large use of social media targeting mainly the left and centre-left, there have been many other attempts to promote more collaborative inputs and joint efforts as well as participatory initiatives. The latter is particularly evident in two large-scale campaigns led by state bureaucrats (law enforcement officials) who received assistance from society representatives to exert pressure on the legislature for the passage of anti-corruption bills. Public prosecutors spearheaded the 10 Measures against Corruption Campaign, while an electoral judge led the *Ficha Limpa* campaign. These bills garnered support from almost two million voters who signed physical petitions (Mattoni and Odilla, 2021). Both initiatives explored social media affordances to raise awareness and mobilise support to gather signatures and exert pressure for the approval of the bills. However, only the *Ficha Limpa* became law. The 10 Measures against Corruption Campaign faced many backlashes and has not yet been passed by the Congresso Nacional (National Congress). Of the two campaigns, *Ficha Limpa* was the most innovative in terms of combining online strategies, including e-petitions, online streaming of events, uploading of hashtags, and overflowing politicians' email boxes, with offline ones such as phone calls and street demonstrations. To a large extent, the 10 Measures against Corruption reproduced the main online strategies previously used by *Ficha Limpa* activists (see Mattoni and Odilla, 2021, for a detailed explanation).

Not only in Brazil but also in Latin America, more broadly democratic innovation has always been characteristically state-driven (Pogrebinschi, 2018). This is particularly important because more traditional scholarship on social movements often depicts state actors as attempting to weaken or co-opt civil society (Holdo, 2019; Piven and Cloward, 1979) or as lacking incentive and adopting a hands-off approach, leaving organising interests on their own, as noted by Rich (2019). These views may sound familiar to anti-corruption scholars who are aware of the hypocrisy in political rhetoric when it comes to promises and real actions to curb corruption (Lagunes et al., 2021a), and of anti-corruption policies being enforced to protect strategic allies and exclude opponents and, hence, to legitimise different types of political domination (Huss, 2020).

The symbolic dimension

The degree of confrontation, as depicted in Table 4.2, dictates the nature of interaction between state and civil society actors, with most favouring collaboration over confrontation. According to Mattoni (2024), the symbolic aspect of ACTs is linked to perceptions of corruption and anti-corruption, influencing a more or less confrontational approach taken to address it. This symbolic dimension also incorporates perceptions of technologies, whether digital or non-digital, reflecting varying levels of enthusiasm and the practical capability to contribute to the fight against corruption. As noted in the previous chapter, in Brazil, integrity techies view ACTs as great allies that are, however, incapable of fighting corruption on their own. Not coincidentally, most of the digital technologies already in place assist humans in anti-corruption efforts.

Evidence of the less confrontational nature of the symbolic dimension of ACTs in Brazil is strengthened by the imaginaries of corruption and anti-corruption that emerged from the interviews undertaken for this study. In Brazil, there is a pervasive perception of corruption, fuelled by several corruption scandals, which seems to have created a demand for action. Tech-savvy individuals mentioned in the interviews that they saw an opportunity or felt the need to use their knowledge and technical skills to react to what they perceived as a widespread issue, combined with a lack of state capacity and political will to curb corruption. The data analysis also suggests that among the participants corruption is widely perceived as the misappropriation of public funds, while anti-corruption is primarily viewed as monitoring. Simultaneously, from the data collected in Brazil it emerged that anti-corruption is seen by many participants as a risky endeavour that is dependent on the state, as illustrated in Figures 4.1a and 4.1b.

Corruption having a multiform nature   Corruption as a sectoral problem
Corruption as business exerting influence on politics
Corruption as abuse of power Corruption as a systemic problem
Corruption as legal corruption Corruption as a moral issue
Corruption due to lack of state capacity in fighting it
Corruption as manipulating procurement (contract, tender, bid)
Corruption as misuse of public money
Corruption as conflict of interest [+]
Corruption as a social phenomenon
Corruption as a problem of laws/rules
Corruption as difficult to identify
Corruption as misuse of public good for private benefit

Figure 4.1a Code clouds for participants' imaginaries of corruption in Brazil
Source: Compiled by the author using MAXQDA.

Anti-Corruption as a risky action

Anti-Corruption not a priority   Anti-Corruption as a time consuming activity

Anti-Corruption as dependent on the State

Anti-corruption as reporting wrongdoings

Anti-Corruption as increasing transparency

Anti-Corruption as monitoring

Anti-Corruption as enabler of democratic participation

Anti-Corruption as increasing civic participation

Anti-Corruption as politicized

Anti-Corruption as a local effort

Anti-Corruption as research and policy advice   Anti-Corruption as a State effort (+)

Anti-corruption as promoting laws developments   **Anti-Corruption as improving integrity**

Figure 4.1b Code clouds for participants' imaginaries of anti-corruption in Brazil
Source: Compiled by the author using MAXQDA.

The risks posed by anti-corruption practices in Brazil are, according to the participants, more related to the risk of being sued for slander or defamation or cancelled online than of being imprisoned. This may happen precisely because there is a close relationship between anti-corruption CSOs and law enforcement agents. Activists see many anti-corruption actions, such as improving the legal framework, releasing data, and investigating and enforcing sanctions as exclusive to the state.

The political dimension

The political dimension is intertwined with the established institutional arrangements within political systems, which frequently delineate responsibilities and constrain the actions of governmental workers, elected representatives, and civil society actors. It is further shaped by political narratives and how reforms are implemented. Consequently, it is directly linked to the design and operationalisation of checks and balances mechanisms, which in turn impact the other dimensions of ACTs. It influences how institutional and civil society actors position themselves and interact in the field of anti-corruption, what kind of digital technologies are in place, and what imaginaries are associated with corruption and anti-corruption, as well as with technologies.

To assess the political dimension of ACTs, it is necessary to map the network of institutions that together form the mechanisms of accountability, as well as the interaction between these institutions responsible for top-down accountability and with actors involved in bottom-up accountability. In the case of Brazil, agencies such as the CGU, the

TCU, the Revenue Service, CADE (*Conselho Administrativo de Defesa Econômica*, or Administrative Council for Economic Defense), the Polícia Federal (Federal Police), the *Ministério Público Federal* (Public Prosecutor's Office), and the judiciary play a law enforcement role in terms of monitoring, investigating, and sanctioning. These agencies represent the three branches at the federal level, and although their formal responsibilities are clearly defined by the legal apparatus, there are important overlaps that result in cooperation and competition (Aranha, 2020). This model is largely replicated at the state level in Brazil. Civil society also has a role to play in this highly complex and competitive environment. In Brazil, civil society responses can be more combative, seeking to punish corruption by, for example, naming and shaming those who engage in misconduct and promoting reputational damage, contributing to the policymaking process, proposing popular legislation and filing lawsuits in the courts to enforce legislation, or even informing voters to "punish" corrupt politicians at the ballot box. However, there is a greater focus on monitoring by civil society actors, due to the perception that there are already other bodies with greater capacity to investigate suspected malpractice and impose corrective measures in proven cases, as well as the dependence on the legislature and executive to improve the legal framework.

The *political dimension* helps to explain, for example, why Brazil does not follow the international trend of having platforms specifically designed for reporting corruption, one of the most popular attempts to use technology to curb corruption by crowdsourcing data on wrongdoings, such as *I Paid a Bribe* (Kukutschka, 2016; Chakraborty, 2024). While Brazilian law enforcement agencies provide a few official channels for submitting reports and complaints, primarily through their *ouvidoria* (ombudsman offices), our study has not identified any widely established platform dedicated solely to receiving such reports; during the COVID-19 pandemic the *Instituto Não Aceito Corrupção* (I Don't Accept Corruption Institute) launched the online platform "Corruptovírus" to receive reports of corruption related to the pandemic, but it had a short life span and it was no longer available online in 2024.

This contrasts with practices in other countries. Italy, for example, offers multiple options, both top-down and bottom-up; for example, the Italian anti-corruption authority partners with a private company to promote *OpenWhistleblowing*. On the other hand, Transparency International and the social enterprise Whistleblowing Solutions have developed *WhistleblowingPA*, a platform offered free of charge that allows Italian public administrations to receive documents from potential whistleblowers and sources in a secure and anonymous way. *WhistleblowingPA* was later renamed *WhistleblowingIT* and expanded to offer paid solutions to private organisations and state-owned companies. Both *OpenWhistleblowing* and *WhistleblowingIT* utilise the free open-source software GlobaLeaks. Meanwhile, in countries such as Uruguay, Peru, and Mexico, media outlets have adopted GlobaLeaks technologies to provide secure channels for reporting corruption.

In India, the *I Paid a Bribe* platform gained international recognition for allowing individuals to report anonymously incidents of bribery and corruption that they have encountered or witnessed (Chakraborty, 2024). The platform aims to raise awareness of the prevalence of corruption across various sectors and to motivate citizens to combat it actively. The information gathered by the platform can identify corruption patterns and advocate for policy changes to mitigate corruption. However, it is essential to note that reporting an incident on the platform does not guarantee any subsequent investigative or legal action against the wrongdoer.

One participant, a civil servant from the CGU, recalled that there had been conversations with CSOs to partner and create digital platforms, but they were not fruitful. Both Transparency International and *Transparência Brasil* considered implementing a landline or online platform to receive denunciations but never moved forward with these plans. There had also been a failed attempt to set up an *I Paid a Bribe* platform in Brazil. One interviewee, a civil servant from the CGU, explained why he thought these attempts had failed:

> So, what happened? I remember that, even with Transparency International, we tried to partner to create those protection-based reporting systems to flow from them to governmental units. It didn't progress ... I tried to help, to create a project, but it's not part of the Brazilian cultural norm to report. And there is the credibility issue. That's one of the things – if we don't provide feedback for those reports, they [the whistleblowers] never inform again. There are apps. We organised a hackathon in Paraíba, and one of the proposals I made was an app so a person could signal, without the need to file a report, a bribe request. So, let's say I'm driving along the road, and there's a highway patrol officer, right? And they asked for a bribe. I'm not going to reveal my identity, I'm not going to report details, I'm just going to note that there was a bribe request was made in that spot. One can check whether other bribe requests were made during that officer's shift. But civil society didn't buy into this idea at the time. This is an app that needs to come from civil society. I think it's crucial ... In Paraíba, we organised two hackathons, and I suggested this topic both times, but it didn't get anywhere.[4]

Interviewees from initiatives such as *Transparência Brasil*, OPS, and OSB recognise that there are important limitations to implementing whistleblowing technologies. Not only are they unable to process and investigate such reports, but also they cannot guarantee punishment for any cases that are proved to be related to corruption. Sanctions are seen as absolutely outside of their capacities. This may also explain why Brazilian grassroots groups have been investing in a more indirect approach to tackling corruption. From the bottom-up perspective, for example, this means improving transparency and access to information, creating new legislation, and carrying out social accountability of public expenditures. There is also a general acknowledgement that certain responsibilities, such as conducting investigations, bringing cases to court, and implementing punitive actions, fall within the purview of law enforcement agents.

The goal of this chapter was to highlight the most significant features emerging from the data analysis. Hence, it is important to acknowledge that ACTs possess unique configurations of material, symbolic, social, and political elements that may diverge from the more general analysis provided by Figure 4.2. While these dimensions and their respective elements are prominent according to the analysis, there may exist other elements within each dimension that warrant consideration. Moreover, it is essential to reiterate that ACTs serve as a heuristic tool, enabling us to move beyond a mere instrumental understanding of digital technologies in the fight against corruption. By incorporating a multifaceted approach that encompasses material, symbolic, social, and political dimensions, ACTs offer a comprehensive framework for understanding and addressing corruption challenges in contemporary society.

Explaining the widespread proliferation of integrity techies and their digital technologies in Brazil

By analysing the four-dimensional characteristics of ACTs, this research has identified at least six elements that could explain the country's innovative role in anti-corruption. First, there is a demand – manifested in corruption scandals that regularly hit the headlines – that reinforces the perception of rampant corruption in the country. As a result, there is both domestic and international pressure to strengthen accountability mechanisms and combat impunity in corruption cases. While corruption scandals permeate all levels of government, including federal, state, and local, the country has gradually moved towards promoting accountability (Power and Taylor, 2011).

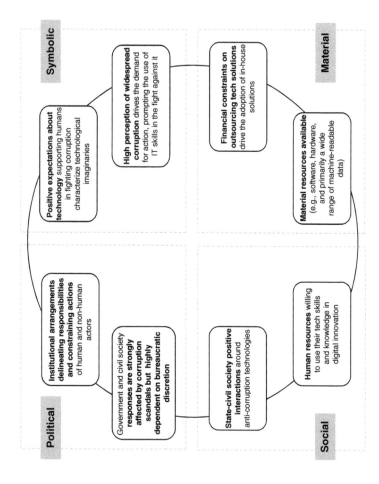

Figure 4.2 The political, symbolic, material, and social elements of ACTs in Brazil
Source: Created by the author.

The second is budgetary requirements. Although it can also be seen as a contextual element, budget constraints also exert an influence, albeit in distinct ways, for governmental and societal initiatives. In the context of law enforcement agencies, factors such as understaffing and constrained budgets have motivated civil servants to devise internal solutions to expedite procedures and streamline their day-to-day operations. It is also noteworthy that a few civil servants from law enforcement agencies who were interviewed mentioned an overall hesitation to share sensitive data with private companies offering digital solutions. One remarkable exception already highlighted by Odilla (2023a) is the Brazilian *Polícia Federal* (Federal Police), which had, in the past, acquired an IBM Watson, a computer system which was originally designed for the quiz show *Jeopardy* but which later was considered a failed ambitious project (Gainty, 2023).

Many bottom-up initiatives have also developed their own solutions, such as Rosie/OSA and OPS; some of the digital tools of *Transparência Brasil* were also designed in-house. While most grassroots initiatives have sprouted from concerned citizens' sense of obligation to take action, budgetary requirements also factor in the proliferation of ACTs. *Transparência Brasil*, for instance, has been developing specific types of digital technologies and projects that are contingent on the availability of grants. One interviewee from *Transparência Brasil* highlighted that donors often prioritise investing in new ACTs rather than supporting those already established, which require maintenance and potential scalability[5]. Many initiators, like the creators of Rosie, initially envisioned generating income from such ACTs, but their struggles to secure funding prompted them to transfer the bot to an NGO.

Third, there has been an effort to combat corruption and improve governance through a more transparent and digitised approach. Consequently, rapid digitalisation processes have made Brazil an extensive open government data provider. As noted by Murray et al. (2023), this foundation is supported by the National Infrastructure of Open Data and the Brazilian Open Data Portal (both established in 2012) as well as the National Open Data Policy (2016). The Open Data Policy introduced structured procedures and accountabilities for releasing open data by the federal executive, intending to ensure consistent availability and reduce vulnerability to shifts in political dynamics. Additionally, the policy strives to harmonise access to information with ombudsman processes. This may also explain the significant number of ACTs related to monitoring and risk detection. Still, numerous challenges persist regarding access to specific types of data. These obstacles stem primarily from the sensitive nature of the data used to fight corruption, the absence of standardised data formats and processing systems, and variations in digitalisation levels across local, regional, and federal governments.

One could argue that bureaucracies often have an interest in keeping their data closed, and wonder why Brazil is any different. The political role of the CGU in promoting transparency and opening up government data has been crucial, despite all the tensions that this process has generated with other state actors. Indeed, the CGU's work in this area has also been driven by CSOs, in particular *Transparência Brasil* and Abraji, which have not only pressed for change but have also actively participated in the design of legislation and the development of ACTs.

Fourth, beyond the availability of digital data, Brazil boasts a pool of integrity techies. The country counts on skilled IT professionals, and, over time, has increasingly embraced innovation. The digital divide between individuals in Brazil is rapidly diminishing, although it is worth noting that digital illiteracy still poses challenges in terms of access to technology (Nishijima et al., 2017; Odilla, 2023a). Furthermore, as discussed in Chapter 3, government agencies have dedicated resources to training and cultivating innovation. This involves governmental initiatives such as organising public competitions to recognise and endorse exemplary practices and technologies, arranging hackathons to collaboratively design anti-corruption prototypes with technologically adept citizens, providing support to government employees interested in pursuing post-graduate courses and establishing innovation labs focused on creating these prototypes, including AI-ACTs (Odilla, 2023a). A kind of spillover effect was also observed, whereby one governmental agency's decision to use emerging technologies to speed up anti-corruption procedures affected how other law enforcement agencies started using their own data to customise ACTs. The same applies to grassroots initiatives, in which former collaborators of bottom-up ACTs created their own digital technologies.

Fifth, there have been institutional efforts in many law enforcement agencies to create networks and share experiences among developers from the civil service and civil society. What initially started as individual initiatives of state bureaucrats was formally embraced by certain agencies that have created laboratories to test ideas and develop solutions. Therefore, elements including decades of accumulated knowledge – often bolstered by state support – and improvements in technology quality and accessibility are likely contributors to the ongoing "AI spring," which characterises the continuous and expanding advancements in the field (Berryhill et al., 2019). In conclusion, access to digital resources and technological expertise are essential prerequisites for AI-ACT development, and Brazil has been making strides in this aspect.

Finally, it is important to highlight the legal framework contributes to the emergence of integrity techies. The legal structure in Brazil includes explicit regulations promoting fiscal transparency, access to information, and the release of open government data in digital formats (Murray et al., 2023), which in turn facilitates grassroots action. Over the years internal control mechanisms and oversight agencies have expanded and digitised thanks to the legal apparatus. For a long time, federal guidelines have regulated public procurement and conflicts of interest, with recent legislation aiming at modernising rules for public expenditures (the 2021 New Public Procurement Law) and monitoring asset declarations and the schedules of public officials (e.g. the 2013 Conflict of Interest Law). Additionally, there are provisions and incentives in place to protect whistleblowers who expose corruption and fraud related to funding, public procurement, and state-owned enterprises (the 2019 Anti-Crime Measures Law). Also, Brazil criminalised money laundering through the 1988 Anti-Money Laundering Act and introduced regulations for anti-corruption compliance in the private sector (the 2013 Clean Company Act, also known as the Anti-Corruption Law). Although we observe a high level of discretion among bureaucrats, which, in the case of law enforcement agencies, has resulted in the acceleration of the digitalisation of anti-corruption efforts, it is crucial to note that the executive and legislative powers still play a significant role in this process that cannot be considered marginal.

Conclusion

This chapter explored the key dimensions of ACTs in Brazil, focusing on their material, social, symbolic, and political dimensions. The Brazilian case draws attention to the development of in-house solutions from both the top-down and bottom-up, emphasising a more monitoring-oriented type of ACTs with a low level of contention among civil society and governmental workers who are part of them. The analysis also identified six elements that, when considered together, may explain Brazil's emergence as a prominent innovator in leveraging emerging digital technologies to assess corruption risks within the public sector.

The Brazilian context is characterised by a demand for anti-corruption measures, driven by a series of scandals and a prevalent perception of widespread and systemic corruption. Simultaneously, technological progress unfolds within a context of financial constraints, where ACTs present a cost-effective avenue for government entities seeking to enhance or replace human endeavours while also offering the prospect of funding for initiatives originating from civil society. Furthermore, Brazil is rapidly undergoing digital transformation, providing accessible public open data and boasting proficient IT professionals within its civil service and engaged citizenries. Ultimately, the country has made significant strides in implementing crucial anti-corruption regulations that facilitate the creation and implementation of ACTs. These elements collectively are believed to have contributed to the country's transition to a position of significance in the realm of anti-corruption efforts. However, as will be explored in the next chapter, some important challenges and limitations may affect the outcomes of ACTs in the country.

Notes

1 For in-depth information on the *Ficha Limpa* (Clean Slate) campaign, see Mattoni and Odilla (2021).
2 See https://buscalai.cgu.gov.br/ (accessed on August 22, 2023).
3 Interview conducted online on December 3, 2021.
4 Interview conducted in Brasília on January 2, 2023.
5 Interview conducted online on December 3, 2021.

5 Outcomes, hurdles, and prospects of anti-corruption technologies

The previous chapters have shed light on the digitalisation of Brazil's anti-corruption efforts, driven by a series of scandals and facilitated by growing public transparency, increased data availability, accessible technologies, and the emergence of actors who are eager to utilise their expertise in the development of digital technologies. These actors, termed in Chapter 3 as "integrity techies," play a pivotal role in the digitalisation of anti-corruption efforts, strengthening public integrity by tackling corruption, promoting transparency, and advocating for accountability in Brazil. The noteworthy skill of most of these integrity techies lies not so much in mastering a specific digital technology but in effectively combining both dedicated and non-dedicated technologies, ranging from low-tech to high-tech, and engaging with diverse data practices. Despite the seemingly innovative and optimistic scenario marked by a proliferation of top-down and bottom-up anti-corruption initiatives that develop and use digital technologies, there are inherent risks and challenges in this landscape.

An important challenge is defining the best metrics to measure the outcomes of anti-corruption technologies (ACTs), which should consider not only the impact on corruption per se but also assessments of its material, social, symbolic, and political dimensions. From the interviews conducted for this study, however, it became clear that both civil servants and activists struggle to measure outcomes. For many of them, the very act of creating an ACT is considered a "success." This is because they still encounter limitations linked to the absence of access to high-quality and standardised machine-readable databases, for example. There are issues in maintaining the operability of ACTs and attracting users. Struggles associated with sustainability and engagement have emerged as critical limitations hindering the enhancement of the accuracy, scope, and effectiveness of both top-down and bottom-up technologies.

DOI: 10.4324/9781003326618-5

The significance of transparent and fair systems raises another pertinent point for discussion: the lack of regulations and norms advocating for auditable anti-corruption digital systems in Brazil. This is particularly relevant for artificial intelligence (AI) applications that are rapidly spreading throughout the country. A very low level of transparency in ACTs has been observed, particularly in governmental ones, and their auditability remains largely confined to their developers. While many bottom-up ACTs have their codes open on GitHub, there is very little incentive to test the fairness of ACT algorithms.

This chapter is dedicated to exploring the main struggles and challenges described above, starting with the limitations concerning the assessment of ACT outcomes. Next, the primary limitations are discussed, as is the necessity to impose regulations on emerging technologies to avoid similar issues that ACTs are designed to address. The governance of ACTs is an urgent topic, but it still is a distant concern among practitioners, policymakers, and academics globally. Yet it cannot be denied that the trajectory of the anti-corruption fight in Brazil is characterised not only by an incremental pace, as analysts have already pointed out, but also by digital innovation. The digitalisation of the anti-corruption fight has positioned Brazil in the vanguard but, at the same time, it should raise concerns for new risks.

Decoding ACT outcomes

What is the impact of the ACTs already in place in Brazil? To what extent are they helping to curb corruption? How can we measure their outcomes? Lessons from the literature on social movement outcomes suggest that we should not limit the discussion to success and failure (Bosi and Uba, 2009). Additionally, those studying social movements outcomes stress that it is never a complete win or lose, as there are many different kinds of effects (Elliott-Negri et al., 2021). To reflect on the outcomes of ACTs, this chapter follows these lessons and uses the goals and expectations of integrity techies as a starting point as well as how they perceive the results of their efforts. It considers that in order to understand what those actors achieved and to evaluate their accomplishments, it is crucial to assess their objectives.

From the interviews conducted for this study it became clear that the simple act of putting an issue related to corruption on the table for consideration and creating technologies to allow monitoring or corruption detection is thought to be a cause to celebrate by many participants. The fact that civil servants, activists, and tech-savvy individuals involved in anti-corruption efforts celebrate raising public awareness of issues such as the misuse of public funds and corruption as a positive outcome may indicate their low expectations. In addition, the findings suggest that there is a struggle to define the best metrics to measure the impact of their initiatives, although statistics such as those related to website or app traffic (e.g. the number of users, new and returning visitors, the length of time that a user is active online) and traffic sources (e.g. social media, direct links, organic search, etc.) are often monitored and used.

Most ACT developers prefer not to limit their metrics to this type of analytics, believing they can mislead the assessment. Activists also consider factors that depend on the type of ACT and their purposes. Social media, for example, is often used to give visibility and to put pressure on decision-makers, sometimes naming and shaming them. In this case, interactions such as the number of "shares" and "likes" may be important. In the case of technologies designed to crowdsource information, such as whistleblowing platforms or applications to receive documentation of monitoring activities, the number of reports received may count, although there is a need to curate this type of information to verify its veracity and quality. Sometimes, outcomes are measured by how the authorities respond, as in the case of ACTs being used to spot suspicious cases.

It is because expectations and outcomes may be very diverse that there is no consensus on which metrics are the best for ACTs. As one Brazilian civil servant[1] noted, one person is sufficient to drive a significant change, as happened many years earlier when he recalls a journalist discovering that a minister had bought a tapioca pudding with a government credit card. The actual cost of the pudding was trivial, but the purchase thereof became a scandal because of the evident lack of boundaries between public and private spending. The case triggered new regulations on the use of government credit cards. This episode invites a discussion of what to expect from ACTs in terms of outcomes and, more importantly, how to measure them.

If we examine the expectations of those interviewed for this research in terms of the goals they wanted to achieve, "willingness to serve society with one's own expertise" emerged as the most recurrent aim in those engaged with both top-down and bottom-up ACTs. While civil servants expressed the goal of increasing the use of ACTs in public administration to suppress staffing shortages and expedite daily procedures, civil society actors voiced the aim of increasing citizen engagement through their tools. While less common than the previously mentioned motivations of serving society with one's own expertise, meeting the demand for ACTs in both the civil service and civil society and engaging people in anti-corruption action through technology, the data also unveiled additional aspirations among participants when they were asked about why they were engaging in anti-corruption efforts. Most of their goals, however, can be categorised into less direct forms of combating corruption, such as aiming to change legislation, ensuring transparency, advocating for open data, and promoting ethical values, and these are not mutually exclusive. Additionally, there are broader goals linked to creating communities to fight corruption, transforming the feeling of powerlessness, and improving people's lives.

It is worth noting the very modest expectations among participants regarding the reduction of corruption and the punishment of corrupt individuals. This reflects their perception of the outcomes of their activities and the ACTs, which are also less closely linked to metrics often used to measure corruption, such as those based on perceptions, corruption experiences, prosecutions, and convictions (Mungiu-Pippidi and Fazekas, 2020). Considering the participants' goals, a typology was abductively created to help analyse the data and that also assisted those who wanted to discuss the outcomes of anti-corruption efforts. Both expected and unexpected outcomes were observed, which can be divided into positive and negative categories. The expected outcomes are aligned with the established goals and corresponded to the initial anticipated projections, both positive and, although less commonly, negative. Unexpected outcomes are unintended and can be both positive and negative, as illustrated in Table 5.1.

As mentioned earlier, most of the observed outcomes of ACTs under analysis have little to do with how many criminal and administrative inquiries were initiated, how many individuals were sent to jail on corruption charges, or even with the achievement of better positions in anti-corruption rankings, the most common existing measurements of corruption. This is so not only because participants have very modest expectations but also because aspiring to a corruption-free country is something considered unrealistic by those creating the ACTs. This is not to say that the expectations of ACTs have nothing to do with reducing corruption. There is a strong belief that increasing transparency, promoting monitoring, and creating awareness are also forms of fighting corruption, although the direct correlation between these activities and reduced corruption is hard to measure through the development and use of digital technologies. The goals of integrity techies do not always align with tangible outcomes once the ACTs have been created and launched. The disparities between expectations and actual results are evident in the fact that unforeseen, both positive and negative, outcomes frequently occur, as illustrated by Table 5.1. In addition, expectations and goals may be adjusted over the years.

Table 5.1 Expected and unexpected outcomes of Brazilian ACTs, measured by their positive and negative impacts

	Outcomes	
	Expected	*Unexpected*
Positive	*Expected positive* For example, helping to recover public money (OSA and OPS); contributing to new legislation (Ficha Limpa/ MCCE, *Transparência Brasil*); contributing to public procurement annulment (Amarribo, OSB), and raising awareness (OSA, OPS, OSB, Amarribo, *Transparência Brasil*).	*Unexpected positive* For example, mobilising hundreds of individuals and creating a community of people with backgrounds in IT, journalists, activists, and concerned citizens on Telegram (OSA and OPS); triggering other bottom-up accountability initiatives (OSA) and expanding to regional and local levels (the TCU and the CGU's ACTs).
Negative	*Expected negative* For example, finding the open resistance of lawmakers to pass new legislation that could be used against them (*Ficha Limpa/MCCE*); threats and intimidation attempts (OSA, OPS), and a limited scope due to the lack of data availability and accessibility.	*Unexpected negative* For example, a small number of users (*Tá de Pé, Esmeralda, Rui Barbot*); the initial resistance of users (Alice, Sisam), and the resistance of governmental agencies to address suspicions and take expected action (*Tá de Pé*, OSA, and OPS).

Source: Compiled by the author.

Operação Serenata de Amor (OSA – Operation Love Serenade) is noteworthy in terms of accounting for expected and unexpected outcomes, as well as readjusting the goals due to different types of challenges faced. Initially, its creators had to convince the Brazilian Lower House (*Câmara dos Deputados*) to make some adjustments in their database to allow a bot to grab all the necessary data necessary to monitor politicians' expenditures. Once they freed up the data and created a digital tool able to spot suspicious cases of the misuse of public money, they experienced the first unexpected negative outcome: they struggled to convince the authorities to act when confronted with the findings of the initiative. OSA's creators did not hide their considerable frustration with the small number of responses from politicians who had misused public money and the law enforcement agents who had encouraged the initiative but did not take action because they considered the findings too insignificant in terms of the misuse of public money; for example, the bot was able to identify irregular expenditures regarding meals and transportation.

Then one of OSA's creators unilaterally decided to use Twitter (now known as X) to publicise every time the bot found a suspicious expenditure. Instead of using a personal account, he created one for the bot, Rosie. This attracted media attention to the initiative due to its novelty and the use of AI. The initiative mobilised 600 people who started interacting in a Telegram group and contributing with ideas and coding voluntarily. This was an unexpected "bitter-sweet" outcome that was both positive and negative. OSA's creators wanted to create an open-source technology but did not expect to have to manage such a big community. Furthermore, politicians started reacting to the Twitter posts; some of them interacted with the bot while others repaid money they had received irregularly, which counts as a positive expected outcome.

Despite the apparent success of OSA, its creators could not attract the necessary funding to employ a full-time team dedicated to the project. Two out of the three initiators left the project, and one of them became responsible for the transition of OSA/Rosie to become part of the portfolio of the non-governmental organisation Open Knowledge Brazil (OKBR), as well as to create new projects – he also eventually left OKBR to work on the promotion of civic tech. During the transition, another unexpected negative outcome occurred for OSA: Twitter declared war against all types of bots and blocked Rosie's account. A campaign was launched to let Rosie tweet again, and the bot came back, but it could no longer tag the politicians' accounts, thereby reducing its visibility and number of interactions. Over the years, it became clear that, even with full automation, there were times when the bot stopped running and/or tweeting due to the lack of maintenance. In January 2024, after over a year without tweeting, a pinned message linked to a crowdfunding website explained that OSA "is not over but it has changed significantly" and "has evolved into other projects," including *Querido Diário* (Dear Diary) and *Perfil Político* (Political Profile) (see the online Appendix[2]).

OSA has inspired other people, including former collaborators and volunteers, to create their own projects, an outcome that was not planned by its founders. One of them was *Dados Abertos de Feira* (Open Data from Feira), which focused on opening data from Feira de Santana, a municipality in Bahia State. The project started in 2018 but had a short lifespan, as OSA also did. *Dados Abertos de Feira* was closed in 2023, even after receiving awards and international funding. The spin-off initiative struggled due to issues accessing public data that was supposed to be open, available, and accessible, according to Brazilian legislation.

Initiatives that account for positive campaigns, such as the one that resulted in the passing of the *Ficha Limpa* (Clean Slate) legislation, led by the *Movimento de Combate à Corrupção* (MCCE – Movement Against Electoral Corruption), had to develop digital and offline pressure strategies not only to have the bill approved by Congress and enacted by the president but also declared constitutional by the Brazilian Supremo Tribunal Federal (Federal Supreme Court). This was necessary due to the open resistance against legislation that bans candidates convicted by a committee for a list of crimes, including corruption, from running in elections. However, the MCCE representatives need to continue monitoring further attempts in the Congresso Nacional (National Congress) to modify and weaken one of the most successful anti-corruption efforts ever seen in Brazil (Mattoni and Odilla, 2021)

From a top-down perspective, some initiatives have also attracted fewer users than expected, which is considered a negative unexpected outcome by all the participants. The bot Alice is often portrayed as a successful story of using AI to predict corruption. However, after much initial hype, the tool designed to send email alerts with suspicious calls before bid evaluation and selection was not used, as expected, by Controladoria Geral da União (CGU – Office of the Comptroller General). Then, the *Tribunal de Contas da União* (TCU – Federal Court of Accounts) showed an interest in the tool. The code was shared and improved, giving Alice a longer lifespan. *Macros*, an online searchable system that acts as a repository for crucial information for inspections and audits (see Chapter 3), can be seen as a case of an unexpected positive outcome as it was initially designed by a civil servant to speed up his work and address the demands of a specific unit within the CGU and, later, it was scaled up and became an official tool of the anti-corruption agency. Not only that, it also started being used in its early versions by other law enforcement agents from different organisations, such as the Polícia Federal (Federal Police) and the *Ministério Público Federal* (Public Prosecutor's Office).

Overall, civil servants prefer to view their technologies as tools that aid human action by automating and accelerating certain procedures instead of relying on technologies to prevent and detect corruption autonomously. Still, the participants recognised that bureaucrats from IT or innovation units, who are more likely to be creating ACTs in the federal government, do not often address their demands. The result is that the use of their technology does not meet the bureaucrats' expectations. Activists and concerned citizens who develop ACTs also report struggling to get citizens to use their tools. The case of the initiative *Tá de Pé*, developed by the non-profit organisation *Transparência Brasil*, illustrates this challenge in trying to convince people to use the ACT to monitor the progress of school construction projects. As mentioned earlier, it began as a mobile app and was later converted into a WhatsApp bot to reach users where they were because the number of downloads and access was very low. Inspired by OSA, they also created an automated bot on Twitter. However, eventually it was discontinued due to a lack of interest.

The present study also found that users' expectations are often not aligned with those of ACT creators. Although the focus of this study is not users' experience and perceptions of ACTs, X (Twitter) followers of Rosie (OSA) and *Tá de Pé* (*Transparência Brasil*) were contacted and interviewed. Their motivations for following these tools varied, ranging from professional reasons to curiosity and a desire to be more involved in bottom-up accountability and anti-corruption efforts. However, they expressed a common level of frustration with the perceived lack of outcomes from both tools in terms of understanding what happened to those named and shamed on Twitter for misusing public money. They expected individuals to be sanctioned, money to be repaid, and legislation to be changed. This was seen as an incentive to use the tools and to continue believing in them. They also expected more in terms of interactions and constant updates. The fact that both initiatives had automated accounts posting the same type of message, changing only the name and state of the politician in the case of OSA and the name of the municipality in the case of *Tá de Pé*, was seen as something repetitive, boring, and in need of improvement.

Overall, the prevailing feeling among followers interviewed was that the tools already in place, while considered interesting and innovative, were not resulting in any profound changes. Both OSA and *Tá de Pé*, as was the case with other tools using Twitter (e.g. Rui Barbot and Zello), were discontinued. The same fate befell the bot Cida, which the CGU had created for Facebook. This adds an unexpected outcome that has not been addressed so far, which is the changes in the guidelines of social media that may create obstacles not foreseen by these initiatives. Often, integrity techies who decide to use these platforms, of which they have no ownership, do so in an attempt to reach a wider audience. However, it often puts them in situations where they lose full control of their intended actions.

Other outcomes are worth noting. At the biographical level of integrity techies, for example, changes could be observed in individuals' career paths after their involvement with ACTs in Brazil. This includes civil servants who initially started using data mining to develop ACTs at the CGU and later left public administration to work in Silicon Valley for major tech companies, tech-savvy individuals who switched to activism and anti-corruption campaign leaders who ran for political office. However, an in-depth analysis of individuals' outcomes was not conducted for this study. At the cultural level, outcomes linked to the emergence of new values and ideas could be observed. This became more salient in the case of anti-corruption initiatives that succeeded in passing the *Ficha Limpa* and contributed to the drafting of the *Lei de Acesso à Informação* (Access to Information Law). The requirement of a clean record has been adopted not only for candidates running for legislative and executive elections but also for federal civil servants in selected positions, as well as for representatives of numerous clubs and associations (Mattoni and Odilla, 2021). The Access to Information Law also transformed how civil society organisations (CSOs) and journalists engage with the public administration and scrutinise its actions, emerging as a pivotal tool for enhancing transparency and accountability not only at the federal level but also in states and municipalities in Brazil.

Addressing four key challenges

The analysis of motivations, expectations, and outcomes from the perspective of integrity techies revealed three key challenges that often undermine the development and use of digital technologies: (1) a lack of accessible and available good-quality data; (2) a low level of users' engagement with ACTs; and (3) sustainability/durability issues. These three challenges were as also observed by Mattoni (2024) when assessing grassroots anti-corruption initiatives across case studies in different countries, including Brazil. A fourth observed challenge emerged during the interviews conducted for this study which is related to the lack of ACT auditability and accountability. Unfortunately, the topic of ethics is still largely neglected by both ACT developers and academics researching the use of emerging technologies, especially AI, in battling corruption, apart from a few exceptions (see Köbis et al., 2022b; Lima and Andrade, 2019; Odilla 2023b). These challenges will now be discussed.

First, a fundamental challenge that has to be faced is how to *deal with problematic data*. Despite all the efforts involved in digitalisation developed to improve anti-corruption action taken in the public sector, as detailed in Chapter 2, the quality of available and accessible data to be used as ACT inputs can still be considered an issue. As we have seen, bad quality data, such as incomplete, inaccurate, or incompatible data, impacts the quality and scope of the outputs of any technology. In addition, there are infrastructural issues related to data quality which can increase the odds of producing skewed results and reinforce existing issues, as well as creating other problems related to bias, noise, and unfairness (Odilla, 2023b).

Although this is a bigger challenge for bottom-up initiatives than for governmental ones, both bodies deal, to a certain extent, with issues that increase the chances of generating errors and limit the scope of the tools. As mentioned earlier, governmental units are more likely to have access to larger datasets, including the ones with data considered to be sensitive. However, when *Macros* was being created at the CGU, for example, the full database of company ownership was not available to its workers. To find basic information about a company, such as its address and whether it was operative, it was necessary to have the unique national register numbers for legal entities and to check them one by one on the *Receita Federal do Brasil* (Brazilian Federal Revenue Service) website, which required a captcha authentication to distinguish between computers and humans. And all this to have access to a certificate that did not include ownership information. Later, the CGU managed to gain full access to the database through a governmental agreement.

A similar issue was faced by the founders of OSA when they attempted to automate the cross-checking of companies that received money from Congress with the database of the Revenue Service. To liberate this specific dataset, two developers and a lawyer joined forces to request access to it through the Access to Information Law, which explicitly states that data should be provided in machine-readable formats. The governmental agency responded by stating that the data was already available online (with captcha), and if they wanted access to the entire database at once, they would need to get it from the company managing the governmental data, which would cost them over BR $ 500,000 (approximately US $100,000). To overcome this issue without resorting to illegal hacking, a collaborative open project was developed to create an application programming interface (API) capable of accessing the information. On GitHub, this API, named *Minha Receita* [3] (My Revenue), has 23 collaborators and features a prototype with a search engine.

Data quality encompasses various aspects that go beyond access. Participants often complained about the lack of consistency, problematic timeliness, and the mixing of irrelevant or complicated information with relevant data. Not only can scraping be problematic due to a lack of standards, but it can also pose challenges for machine learning (ML) models as they may struggle to learn from these types of datasets. Statistical techniques are often used as an attempt to mitigate access to problematic data. Another layer of complexity arose with data privacy legislation. Many important datasets, such as affiliation to public parties and city property ownership registers, are no longer fully available online, with the argument that they contain confidential information that needs protection. Problematic data, therefore, can hinder decision-making and analysis, and Brazil continues to face this issue.

Second, *how to engage users* is a significant challenge that both civil servants and activists constantly need to address. ACTs are often designed to be used by diverse groups of people, each with their own perspectives, priorities, and communication preferences. Finding common ground or effective communication strategies that resonate with a wide range of individuals is something that participants mentioned they struggle with. Users also expect more in terms of communication of impact, as discussed earlier in this chapter.

This suggests that there is a need to pay more attention to the user's experience by staying in touch with them and collecting permanent feedback; however, this was often neglected by the creators of ACTs, as was observed during the interviews conducted for this study. It seems that users expect a level of interaction that is not only informative but also inspires action or involvement. This is linked to trust and credibility. If a tool is not perceived as effective, i.e. it does not produce the expected outcomes, thereby offering false positives or overloading humans, it may create obstacles to its use. Generally developed to speed up procedures and help individuals in anti-corruption efforts, developers in public administration agree that sometimes their solutions may overwhelm their colleagues as machines are much faster than humans. This was also observed in a bottom-up initiative that sought to cross-check the expenditures of members of parliament. The "red flags" raised by Rosie overloaded the employees working in the department responsible for identifying irregularities and cancelling the reimbursements.

In addition, considering that in a highly digitalised society people are bombarded with information from various sources, cutting through the noise and capturing users' attention is a topic of permanent discussion among developers. Using social media, especially commercial platforms where people are already active, has been seen as a solution to engage with a wider public. This applies to both top-down ACTs attempting to interact with citizens and bottom-up ones, which are often designed to reach a broader audience. However, relying on these commercial platforms also imposes challenges. ACT' actions can be limited due to the guidelines and functionalities of, for example, Facebook, Twitter (now X), YouTube, Discord, Telegram, or WhatsApp. The governmental bots Zello (TCU) and Cida (CGU) stopped being operative on Twitter and Facebook due to changes in the platforms. The grassroots bots Rosie and RobOPS also faced restrictions on Twitter and were blocked for not complying with the platform rules, requiring their creators to negotiate with Twitter to convince them that they were civic bots doing good. *Tá de Pé* had to discover how to circumvent certain restrictions imposed by WhatsApp to automate the informal dialogue and ask people to monitor school construction sites in person. It is worth mentioning that the visual representation of all these bots was charmingly childlike, and most of them had people's names to encourage familiarity and resonate with social media users.

Another way to attract users' attention mentioned by the participants is adding gaming mechanisms into non-game technologies such as ACTs. Although less common in public administration, gamification has been observed in one of the ACTs developed by the TCU. Agata was designed to use human action to train algorithms by applying an active ML process based on the keywords most searched for by auditors. Agata asks the auditor to say whether the result was what the person was looking for and learns by considering the context in which the word is being used. The process was constructed as a game in which the user gained agates (agate stones) to engage auditors, according to one interviewee.[4] At *Transparência Brasil* there had been a discussion about gamifying *Tá de Pé* by creating rewarding activities, such as shareable rankings of those monitoring the most construction sites, and even financial incentives. This particular individual, an interviewee[5] for this study, believes that gamification is the best way to provide a more satisfactory user experience and to incentivise users to stay engaged longer in the case of bottom-up accountability ACTs. However, the organisation decided not to introduce gaming elements in its mobile app designed to monitor the delayed construction of schools.

Transparência Brasil opted to focus on reaching out to activists, who are already engaged citizens, from other sectors to help them to improve public integrity and accountability using ACTs in their areas of interest. Law enforcement agencies believe that they also need a better dialogue with CSOs to increase the use of the open data. At the CGU, the Brazilian anti-corruption agency at the federal executive branch, law enforcement officials are already considering co-developing tools that can be used by CSOs. Thus, there seems to be consensus among government officials and CSOs interviewed for this study that there is a need for intermediaries to structure digital anti-corruption efforts.

This is an important finding that emerged from the data that the literature often does not discuss. Not everyone is interested in engaging in anti-corruption activities, especially if they are perceived as being too time-consuming, for example, monitoring and pressuring politicians to pass anti-corruption legislation, or being too risky, such as leaking documents or reporting wrongdoings. Even if Brazil had seen massive anti-corruption street demonstrations over the years, persuading thousands of people to join public protests is not something that happens every day. A high level of disappointment could be expected among both civil servants and activists after realising that ordinary citizens do not naturally engage with data and ACTs to fight corruption.

Third, there is a key challenge related to addressing the *short life cycle of ACTs*. Although planned obsolescence is not a common feature, most of these technologies are likely to have short lifespans due to a combination of lack of use and the disengagement of their creators who have moved on to attempt to develop something else, not always related to anti-corruption. Before any cycle of upgrades or attempts to make ACTs more integrative can occur, they are simply discontinued. Sometimes, it is even difficult to find documentation for them.

One exception regarding documentation is the app *Mudamos*, created to utilise blockchain technology and to offer a secure platform for any citizen to propose draft bills and collect verified signatures to support the proposals. Launched in March 2017, the app rapidly received thousands of draft bill proposals and signatures and managed to discuss hundreds of new legislations in eight legislative houses. In an attempt to improve its usability, an updated version 2.0 was launched, and its creators, the MCCE and the Institute for Technology and Society intensified its communication strategy. However, it can be said that the app faced a low adoption rate (350,000 active users) in relation to the number of signatures required to propose a national-level draft bill (1.5 million) at the Brazilian National Congress. Hence, in 2023, the app was suspended. The code was then opened on GitHub, and instead of offering a service for citizens to propose new legislation, *Mudamos* started offering the system to organisations looking for a secure and affordable electronic signature system protected by blockchain technology. The history of *Mudamos* is documented in the form of a timeline on its website.[6] OSA, on the other hand, took over a year to post on Twitter to say that Rosie had stopped tweeting, but did not document why exactly the project was not being prioritised anymore. Due to its high level of automation, the dashboard Jarbas was kept operative and nothing else.

Both cases exemplify the short life cycle of ACTs, even in the cases of those implementing emerging technologies such as blockchain and AI. One possible explanation for their short life cycle is the high level of experimentation without precise calculation of demand forecasting. According to the participants in this study, many ACTs are created without the real demands and interests of users being assessed, making them more likely to become never-implemented prototypes. As stressed by one interviewee,[7] this holds true for bottom-up tools presented, and even awarded in hackathons, for example, as well as for tools developed in the public administration for the promotion of innovation policies. Insufficient funding for promoting, enhancing, and scaling ACTs, along with a shortage of high-quality data to expand their scope, also undermine the durability of these tools. Furthermore, the fast-paced evolution of technology can lead to the perception that existing ACTs fall behind in terms of functionality and impact, making them appear less attractive to users.

Sustained investment is essential to ensure the continued use and maintenance of ACTs. Many initiatives analysed for this study have been discontinued due to insufficient funding and/or a lack of use. To prevent operational breakdowns, resources, coordination, and trust are required among users and tool developers. The sustainability of these technologies is closely intertwined with the political climate of governments, which can act to interrupt, dismantle, or fail to support both bottom-up and top-down initiatives. It is not a coincidence that many activists are concerned about threats and legal action related to their work once such technologies potentially increase the scope and visibility of anti-corruption practices. Civil servants also face the risk of jeopardising their careers if they persist in developing or using specific solutions without adequate political support from the administration, as was illustrated by the experiences of Jorge Jambreiro Filho (see Chapter 3).

This study has also revealed how both CSOs and the civil service struggle to retain individuals involved in creating ACTs. Many of them are drawn to other projects, sometimes beyond their respective organisations. In the case of OSA, two of the three founders moved on from the initiative to pursue roles with a more commercial focus. The third founder oversaw the transition of the bot to Open Knowledge Brazil for a period and later became involved in civic tech work with the World Bank and then returned to the traditional IT job market, thus following a similar path. Among civil servants, it is common to witness developers from the CGU migrating to other agencies, such as the TCU, where working conditions are considered better, or even moving to Silicon Valley. Notably, the former head of the CGU's *Observatório da Despesa Pública* (Public Expenditure Observatory) left to work for Meta (formerly Facebook), and another member of the same unit left the civil service to work as a data scientist for a fitness and diet app.

The fourth challenge to be addressed is the *ethical challenge* of ACTs. This challenge has been largely unexplored by those researching corruption and has not been openly discussed by those who create and deploy emerging technology in anti-corruption efforts when compared to the use of predictive tools in intelligence, counterterrorism, and policing activities, for example (for a critical view from these other areas, see Angwin et al., 2016; Jefferson, 2018; Siegel, 2018; Edler-Duarte, 2021). There are lessons to be learned from these other fields, which have been documenting misleading outputs, bias, and the unfairness of predictive data analytics while they gain popularity in surveillance, crime, and criminal justice actions (Odilla, 2023b). Furthermore, there are scandals related to AI-based ACTs that remain to be further assessed in order to understand how to address ethical challenges in the digitalisation anti-corruption efforts.

Many governments are abandoning or are being forced to ban the use of AI to combat fraud and corruption. For example, a Dutch court invalidated a welfare fraud detection system that used personal data from different sources for not complying with the right to privacy under the European Convention of Human Rights (van Bekkum and Borgesius, 2021). In China, the AI system called Zero Trust was created to analyse extensive datasets to evaluate the job performance and personal characteristics of numerous government personnel, including information about their assets (Aarvik, 2019; Chen, 2019). However, since 2019, this system has been discontinued in numerous counties and cities, allegedly due to concerns regarding the potential occurrence of false positives and an imbalanced effectiveness in detecting specific wrongdoings (Chen, 2019).

Other scandals have led to investigations and the issuance of numerous recommendations, such as the Robodebt case in Australia (Knaus and Henriques-Gomes, 2023). The country replaced the former manual system of calculating overpayments with an automated data-matching process to detect discrepancies between fortnightly income reported to the unit responsible for delivering social benefit payments and information held by the tax office. This comparison has been criticised as too "crude" (Bucci, 2023). A whistleblower accused the government of retrieving debts from hundreds of thousands of Australia's lowest-paid and most vulnerable citizens and being overly punitive to those on sickness benefits (Knaus and Henriques-Gomes, 2023).

In Brazil, very little is known about potential issues related to ACTs that utilise emerging technologies. *Transparência Brasil* took the first step towards mapping the use of AI in public administration more broadly with an online catalogue, and Odilla (2023a) made an initial attempt to list ACTs. A recent audit by the Brazilian Federal Court of

Accounts in 2022 revealed that 62% of the 263 agencies in the public administration are either in the process of implementing or have already implemented some form of AI system, each at varying levels of maturity (TCU, 2022). While 50% of these agencies developed their tools in-house, the majority acknowledge a shortage of skilled personnel to develop and utilise such technologies (TCU, 2022). However, overall, there is still a lack of comprehensive information on the use of emerging technologies.

The absence of an open repository where both bottom-up and top-down ACTs and their main features, including the technology deployed, can be found reflects the significant opacity surrounding the use of emerging technologies. Providing transparency for these ACTs is an essential first step. This would not only facilitate necessary attempts to assess them but also provide valuable information for users and others interested in developing new systems. Another important challenge is to introduce clear guidelines that mandate the inclusion of mitigation measures for potential bias, noise, and unfairness when creating ACTs, as well as rules that allow for their auditability. Even in the case of technologies dealing with sensitive data, external oversight is desirable and can be achieved through non-disclosure agreements. Having open debates on ethical challenges is also an important step that needs to be taken. The participants noted that when creating ACTs they made efforts to find statistical solutions to reduce issues such as sample bias and incomplete data. However, they recognised that discussions on algorithm ethics and accountability were not often held. As noted by one of the participants, a civil servant from the CGU, not everyone wants to "face criticism" but it is necessary "to know what is needed to be improved."[8] Indeed, not all governments are prepared to undergo scrutiny due to the technologies to which they have contributed or the data they have opened and made accessible to the public.

Conclusion

This chapter shed light on the significant limitations and challenges encountered in the digitalisation of anti-corruption efforts in Brazil. Despite the optimism surrounding innovative developments in the country, the inherent obstacles should be acknowledged, particularly given the rapid evolution of technologies such as AI. Key issues encompass the need to establish metrics for assessing the efficacy of ACTs, the limited access to standardised, machine-readable data, and the challenges in engaging users with existing ACTs. Furthermore, the absence of auditability in many deployed ACTs presents risks of unfairness, bias, and noise in the battle against corruption. Urgent solutions are required in various domains, including infrastructure, social engagement, sustainability, and ethics.

Regulatory frameworks have the potential to strengthen the deployment, integrity, and effectiveness of ACTs, particularly by addressing the need for higher-quality data and the establishment of auditable anti-corruption digital systems, all of which are significant concerns in Brazil. Even if a government has not yet passed dedicated legislation on the topic, as is the case in Brazil, there are many models of digital governance available which may be adopted or customised (Filgueiras, 2023). As Filgueiras and Almeida (2021, p. 4) have stressed, institutions such as the United Nations and the Organisation for Economic Co-operation and Development have been providing broad and diverse guidelines and models for regulating the digital transformation process that can be compiled and incorporated into practice by governments and agents.

However, regulations alone may not suffice to address all the challenges identified in the Brazilian context. Effectively addressing these key issues in the digitalisation of anti-corruption requires a holistic understanding of government, policy, public engagement, and law enforcement. In addition, the dichotomy between bottom-up and top-down ACTs cannot be the sole design consideration if the aim is to develop more effective, enduring, and ethical technologies. Alternative solutions, such as involving civil society in holding governmental ACTs accountable while exercising due caution due to their sensitive nature and fostering collaborative efforts between civil society and government to create new tools, can serve as a starting point for transcending this dichotomy and pooling resources to tackle common challenges.

Notes

1 Interview conducted in Brasília on January 5, 2023.
2 See F. Odilla (2024). *Anti-Corruption Technologies in Brazil: Online Appendix*, available at https://osf.io/7gzwu/.
3 See https://github.com/cuducos/minha-receita?tab=readme-ov-file; https://docs.minhareceita.org/ and https://medium.com/serenata/o-dia-que-a-receita-nos-mandou-pagar-r-500-mil-para-ter-dados-p%C3%BAblico s-8e18438f3076 (accessed on February 2, 2024).
4 Interview conducted online on February 2, 2021.
5 Interview conducted online on December 15, 2021.
6 See https://www.mudamos.org/ (accessed on February 11, 2024).
7 Interview conducted in Brasília on January 2, 2023.
8 Interview conducted in Brasília on January 2, 2023.

6   Conclusion

What kind of anti-corruption technologies (ACTs) do we need? The easy, utopian, general answer would be those that are efficient in detecting, preventing, and reducing as many types of corruption as possible; technologies that are explainable, auditable, and fair; that do not produce biased or noisy outcomes that compromise their quality; that are widely used by large numbers of people; and, most of all, that are durable. Do we have these technologies already? Not yet. What do we have now? The case of Brazil explored in this monograph helps to address this question more broadly. It takes a hopeful perspective, albeit with a grain of salt.

As detailed in the book, Brazil's journey in the fight against corruption combines incremental progress in accountability mechanisms and digital innovation, positioning the country at the forefront of the digitalisation of anti-corruption efforts. Interestingly, in Brazil, this journey did not commence as one might expect, being less about having a strategic vision, investment in emerging technologies, or an existing culture of innovation and collaboration. The case of the digitalisation of anti-corruption in Brazil is more about experimentation and attempts to address corruption scandals, prompted by the general perception that corruption is a widespread systemic issue. Instead of expensive outsourced projects, there has been a proliferation of in-house solutions, facilitated by large amounts of pre-existing digitised data. These data are the result of an earlier process of digitalisation of public finances in the country, which intensified in the 1990s, partly in an attempt to improve accountability.

In the case of top-down ACTs, the process of digitalisation of anti-corruption also represents an attempt to automate tasks and expedite procedures due to limited staffing and the need to speed up procedures. For civil society, innovating to combat corruption often depends not only on the availability of funding for organisations investing in technologies but also on digital literacy and openness to digital innovation among leaders and supporters (Odilla, 2024). In addition, the role of the state in creating digital

DOI: 10.4324/9781003326618-6

databases, opening up data, improving transparency and access to information, encouraging civil society action, and promoting innovation has been crucial in creating a positive environment for a wide range of ACTs in Brazil. The country's federal anti-corruption agency, the *Controladoria Geral da União* (CGU – Office of the Comptroller General), played a key role in this regard, along with civil society organisations (CSOs). In this environment, once ACTs are created and become known for their innovative features, a spillover effect of similar tools is observed, although there is little documentation of what was created by whom and why.

One of the main reasons why the case of Brazil is noteworthy is that civil servants and concerned citizens are developing their own digital solutions with very little outsourcing. However, when this monograph was finalised in February 2024, both the CGU and the *Tribunal de Contas da União* (TCU – Federal Court of Accounts) were conducting studies and holding initial internal discussions to open up their innovation plans and collaborate with start-ups. TCU's experience with ChatTCU, outsourced from a consortium of startups, may be the first step in a new more collaborative trend between the public and private sectors in terms of developing new anti-corruption tools using emerging technologies, as is the case of generative AI. As demonstrated in the previous chapter, civil servants have also been considering how to involve activists and CSOs in co-creation to use public open data more efficiently. In the CGU, there have also been attempts to customise solutions within larger software and cloud packages that have already been contracted, as there is a perception that the market is moving faster in terms of innovation.

These initial discussions to open up anti-corruption innovation plans could reconfigure the current version of "integrity techies," i.e. governmental and civil society actors who shape and are shaped by digital technologies and who influence the social and political aspects of public integrity. In both top-down and bottom-up initiatives in Brazil, so far integrity techies include individuals, their organisations, and collective actions that are directly linked to digital technologies deployed to fight corruption. Among the individual motivations that emerged in the interviews held with those who created or facilitated the development of digital technologies in anti-corruption was a willingness to use personal skills to make a positive impact in society. Additionally, participants noticed that this can enhance one's curriculum vitae and GitHub profile, especially for tech-savvy individuals, and this was mentioned as a strong motivation to engage, particularly in open-source initiatives. Many developers have left anti-corruption projects to work or continue to work in the technology industry. In some cases, civil servants have left the public administration to work in Silicon Valley.

Government integrity techies interviewed for this study are predominantly men who have passed formal exams requiring the candidate to hold at least an undergraduate degree, and they often enjoy higher salaries and job stability within law enforcement agencies. Civil society integrity techies interviewed are more likely to be gender-diverse and aim to make a living from ACTs, but often face challenges in obtaining funding and work solely on their own projects. However, further research is needed to better understand the demographics of integrity techies, as well as the details of their personal incentives and biographies, in order to identify clear differences between government and civil society integrity techies beyond those noted in this book. The same goes for integrity techies who develop digital anti-corruption solutions for corporate use and for those using ACTs.

While broadening the comparison between government and civil society initiatives that use technology to fight corruption is a necessary new avenue for research, there were notable differences found in Brazil. Until 2024, top-down approaches have been more related to "closed innovation", as they follow a model where anti-corruption technical solutions are more likely to be generated and refined in-house without or with very little outsourcing, and linked to the duties of specific agencies due to their law enforcement responsibilities and protective attitude to sensitive data. Bottom-up anti-corruption approaches are often open-source and developed, or at least improved or updated, in a more collaborative vein. Overall, the convergence of social and political aspects of public integrity within the realm of technology delves into practices related to combating corruption, improving transparency, and enhancing accountability within governmental institutions and among public officials.

As discussed in Chapter 4, when the symbolic, material, social, and political dimensions of ACTs were examined in order to analyse the digitisation processes of anti-corruption in Brazil, social perceptions and expectations about corruption, anti-corruption and technology (symbolic) and the availability of digital technologies, including data, software and hardware (material) were crucial. Interactions (social) between state and civil society actors are not only a relevant part of ACTs, but in the Brazilian case they have been essential for the implementation of many bottom-up initiatives. Most of them were facilitated by state action, in particular by law enforcement officials from anti-corruption agencies (mainly the TCU and the CGU). The existing institutional arrangements in political systems, which delineate responsibilities and constrain the actions of different anti-corruption actors, are also an important dimension of ACTs. Therefore, this volume complements Mattoni's (2024) framework for ACTs by adding a fourth key element of ACTs: the political dimension.

Concerning these four elements of ACTs, there are important components that require attention. On the *material* side, we should pay attention to the levels of technological ownership, i.e. the extent of reliance on commercial media as opposed to dedicated tools created to fight corruption. Currently, in Brazil, we have a complex ecosystem in this regard. We observed cases of mixing existing social media and new technologies, the use of commercial platforms to reach more users, or even their marginal use by CSOs that either value face-to-face activities or were unable to catch up with the rapid technological development, as documented by Odilla (2024). On the *symbolic* side, contentiousness is at stake. Most of the ACTs in place have digital technologies designed to expose corruption, which can be done on higher or lower levels of contention. For example, ACTs that allow monitoring often tend to be less contentious for highlighting risks or "red flags" than those designed to publicly name and shame suspicious cases or expose corruption through leaks.

The *social* aspect of ACTs can be observed not only in the type of state-civil society interactions but also in their collectiveness, as some tools depend more on individual users than others. While the *Tá de Pé* initiative requires individuals to use their mobile telephones to conduct monitoring, the bot, Rosie, boasts a higher level of autonomy since its outcomes are available to anyone with internet access, despite relying on a call for users to follow X (formerly Twitter). The *political* element explains why certain actors, especially governments, invest in digital technologies to fight certain types of corruption and helps us to understand institutional and civil society responses to corruption scandals. Overall, considering the "web of accountability" institutions and the accountability cycle, most existing ACTs in Brazil have oversight functions, some support investigations and can be used to pressure for sanctions, but enforcing corrective action is not part of their main function.

Frontiers of ACTs in the automated society

Although the case of Brazil has specificities, the empirical evidence gathered for this study allows us to make assumptions and generalise further about what to expect from ACTs. This is because Brazil has been deploying different types of ACTs designed to fight different types of corruption using technologies ranging from transparency portals and mobile web applications to artificial intelligence and blockchain. In addition, both types of digital technologies deployed in Brazil and type of corruption they were designed to fight against are not specific to the country. Issues related to public spending with public procurement or with elected representatives, delays in public services such as the construction of schools and the purchase of school meals, judicial slowness, cartel practices, customs fraud, the lack of adequate channels to report corruption, and creating new anti-corruption legislation are not problems unique to Brazil or even to the Global South.

Furthermore, the limitations, risks, and both anticipated and unanticipated outcomes observed in Brazil are also not unique and can be readily found elsewhere. The still-limited access to standardised, machine-readable quality data and difficulties in engaging users with the ACTs in place are prevalent almost everywhere. Mattoni (2024) also identified the same three limitations found by this study while examining empirical evidence from diverse bottom-up anti-corruption initiatives across multiple countries, among them Brazil. Opacity, including the scarcity of documentation regarding inputs, data processing, and outputs of most of the ACTs deployed, is an important theme that emerged from the data collected in Brazil and analysed for this monograph. Moreover, the short lifespan of ACTs has not been well documented by those responsible for creating or maintaining digital technologies. All of these increase the risks related to the development and use of technologies in the fight against corruption and compromise the positive outcomes of such technologies. Solutions such as better metrics, sustainability, engagement, and regulatory frameworks to fortify the use of ACTs, their integrity and effectiveness in combating corruption are necessary everywhere.

Given the rapid advancement across various technological domains and the widespread automation within societies, addressing these infrastructural, social, sustainability, and ethical challenges becomes imperative for the development of more decisive and less problematic ACTs. This is particularly valid for cases where the digitalisation of the anti-corruption fight has developed faster than the instruments necessary to measure its impacts and hold them accountable. From the data collected and analysed, we cannot deny that, by now, institutional processes, grounded in the form of regulations and legislation, are equally important in both combating corruption and creating the necessary environment for conceiving ACTs as instruments of public policy and accountability.

In this sense, the "instrumentation of public policy" as introduced by Lascoumes and Le Galès (2007) is helpful for critically reflecting on integrity techies and their digital technologies. This is because, according to the authors, certain public policy and accountability instruments are viewed as a fundamentally technical approach to problem-solving, and are expected to be naturally more efficient and neutral, which they are not. However, with few exceptions, the perception among many practitioners and academics prevails that technologies arrived to solve the problem of corruption. The observations in this study instead urge caution regarding tech determinism and act as a reminder that technology development and its use is also a political choice.

First, digital technologies are neither neutral nor are they often more efficient than humans. Technologies are not neutral because they often grapple with problematic data and rely on historical data with the presumption that patterns will recur to some extent. Those developing digital solutions to fight corruption also have values and beliefs, and as our study suggests, often reflect little on the risks and limitations of the tools they are creating. Hence, considering the existence of problematic inquiries and sanctions, problems may be perpetuated if no attempt at open discussion is made to identify past issues and attempt to mitigate them.

Second, technologies are not always more efficient or immune to fatigue; online systems can suffer from bugs, lack of memory, and various other glitches that compromise outcomes. Yet they can overload humans with information in an already overwhelmed and datafied society (Couldry, 2012), and instead of facilitating human action they can create inaction and distrust. This is more likely if technologies offer unattractive data or false positives or are not perceived as useful to investigate and punish corrupt individuals and improve the quality of government.

In addition, digital anti-corruption solutions have been created and used without a devoted analysis of their relevance and an evaluation of their effects. Ongoing research and open discussion shedding light on each type of technology, its outcomes and risks are also insufficient. Similar to how other researchers have scrutinised facial recognition and risk assessment algorithms linked to various applications like credit scoring and judicial decisions, it should be applied the same level of inquiry, which has already found serious issues related to bias, noise, and unfairness, to technologies aimed at detecting and preventing corruption. Table 6.1 summarises the lessons learned from Brazil in this regard that can be helpful in other countries.

As illustrated in Table 6.1, each type of technology possesses distinctive features that should be considered when assessing ACTs. While the specifics of their risks and limitations may differ, common concerns must be addressed regarding infrastructural aspects, ethics, and user engagement. The first two can be addressed via legislation to increase the quality and accessibility of data, which is the basic raw material for any type of digital

Table 6.1 Types of technology used in anti-corruption: features, risks, and examples in Brazil

Type of technology	Features observed in the accountably cycle	Limitations and potential risks	Examples of Brazilian ACTs
Transparency portals and open (big) data	Digital architectures designed to store large amount of systematised public data, enhancing active transparency and accountability processes mainly through monitoring and investigating. They can be open to anyone with access to the internet or closed for internal use only.	The absence of standardised data and incomplete and complex information, alongside non-machine-readable data, present risks of inaccurate analysis and impede engagement with a wider audience.	*Portal da Transparência*/CGU; *Fiscobras*/TCU; *Dados Abertos*/Lower House; *Portal de Compras do Governo*; Federal/Federal Executive; *e-proc*/Federal Judiciary; *Fala.BR*/CGU; *Dados Abertos do Governo Federal*/CGU; *Base dos Dados*
Bots for monitoring, risk detection, and integrity promotion	AI-based software programs, most of which run online, designed to quickly gather and organise vast amounts of data, perform analytics, create rankings, raise red flags, and send reminders related to risks, audits, and inspections.	The lack of compatible systems and standardised data, incomplete information, and data security are concerns. These systems are often run autonomously with very low levels of accountability and auditability, posing a risk of misleading analysis, bias, noise, and unfair outputs.	*ContÁgil*, Sisam, Aniita/Revenue Service Alice/CGU and TCU Monica, Agata, Carina, Adele/TCU Rosie/OSA-OKBR

Type of technology	Features observed in the accountability cycle	Limitations and potential risks	Examples of Brazilian ACTs
Chatbots	Computer programs are designed to simulate conversation with human users, especially over the internet on social media, to crowdsource information and/or provide (anti-) corruption-related documents.	Audiences are limited, as they can be used only by those on specific platforms. The risk of being discontinued is due to changes in social media internal guidelines.	Zello/TCU Cida/CGU *Edu/Transparência Brasil* Rosie/OSA-OKBR Rui Barbot/JOTA RoboOPS/OPS
Desktop and mobile oversight applications	Digital architecture designed to monitor the performance of public officials and promote transparency with functionalities such as filters, maps or ranks using open public data.	Limited access to data, issues to increase the scope reach a broader audience and convince people to use the tools.	*Políticos do Brasil, As Claras, Excelências*, and *Deu no Jornal/Transparência Brasil* *Perfil Político*/OSA-OKBR *Corruptômetro*
Social media	Interactive commercial technologies that enable the generation, sharing, and consolidation of content, ideas, and interests.	Actions are restricted by the guidelines and changes of commercial platforms and only reach users of certain outlets. Risks include overlooking marginalised voices, contributing to polarisation and fostering misinformation.	OSA/OKBR RoboOPS/OPS *Tá de Pé/Transparência Brasil* Rui Barbot/JOTA Zello/TCU Cida/CGU
Blockchain	Distributed ledger systems to replicate and disperse transactions among a network of devices, thereby enhancing security. They enable reporting of corruption cases, gathering signatures to support anti-corruption bills, or sharing information on investigations more securely.	Potential anonymity facilitates false charges and accusations, technical vulnerabilities undermining data integrity and challenges in reaching a broader audience due to digital literacy gaps.	*Mudamos* app/MCCE and ITS

Type of technology	Features observed in the accountably cycle	Limitations and potential risks	Examples of Brazilian ACTs
Crowdsource online platforms	Electronic systems that harness the collective efforts and resources of individuals to address tasks, solve problems, and generate content related to reporting wrongdoings, raising awareness, monitoring governmental actions, and participating in decision-making.	Technical vulnerabilities undermine data integrity and challenges in reaching a broader audience due to digital literacy gaps.	*Tá de Pé Obras/Transparência Brasil*
Internet of Things	The interconnection of everyday objects' computing devices via the internet enables data transmission and reception to facilitate informing, raising awareness, and receiving information on corruption, particularly when passing by locations of investigations or monitoring activities.	Privacy breaches, data manipulation, the spread of misinformation, and reliance on outdated information, alongside challenges in reaching a broader audience due to digital literacy gaps, are key concerns.	Not identified

Source: Compiled by the author.

Note: For detailed information on the ACTs, see Chapters 3 and 4 in this volume, Odilla (2023a), and the online Appendix available at https://osf.io/7gzwu/.

Caption: Table with four columns and eight rows. Each row provides detailed information on one type of technology, along with their respective features when applied to anti-corruption, limitations, risks, and examples used in this book.

technology, and to introduce measures to mitigate certain types of risks. A regulatory framework, despite impacting ethical and infrastructural risks, is not enough to reduce them to desirable levels. In addition, regulation is not sufficient to improve user engagement.

Reconfiguring governance and updating the social contract

When considering the feasibility of implementing digital technologies where corruption is systemic, as is the case in countries like Brazil, one is inevitably faced with a fundamental dilemma. Can such advances really flourish in such environments of systemic corruption, or even where corruption exists but is not perceived as a major problem? One wonders whether we run the risk of adopting a technocratic approach to a deeply rooted socio-political problem. And even if all the necessary instruments were meticulously put in place, what if there were widespread apathy towards their use? This scenario echoes the current situation in Brazil, where despite the availability of several digital anti-corruption solutions, they are not being used to their full potential and may not be perceived to be provoking much change in the status quo. It is, therefore, necessary to go beyond the mere existence of technological solutions when evaluating ACTs, which must be considered as assemblages of material, symbolic, social, and political elements.

To enhance the potential of ACTs to endure and aid the anti-corruption effort more sustainably, governance may have to be reconfigured for the digital realm. This reconfiguration should not lean towards an expansion of either the state or the market (Filgueiras and Almeida, 2021), but rather towards the establishment of a "new social contract" aimed at improving the relationship between individuals and governing authorities, which would work as an indirect approach to anti-corruption, as proposed by Bo Rothstein (2021). Social contract theory, nearly as old as philosophy itself and key in the work of thinkers such as Socrates, Thomas Hobbes, John Locke, and Jean-Jacques Rousseau, posits that individuals' moral and/or political obligations are contingent upon an agreement among them to establish the society in which they live (Friend, 2024). This agreement, or contract, exists between the ruled and their rulers, defining the rights and duties of each. Rothstein (2021) argues that a functioning social contract is essential for controlling corruption in all societies. Although the specific content of this social contract must be tailored to the unique conditions of each society, it implies that citizens, as one party to the social contract, have the right to demand a "clean government" from the other party (Rothstein, 2021).

Having new clear guidelines to enhance the quality and durability of ACTs is undeniably imperative, and we can work on that. However, focusing on both the development and the use of digital technologies for better governance is equally crucial, ensuring that they are founded on principles such as respect for human rights, adherence to international law, and the provision of meaningful opportunities for all individuals and nations (Filgueiras and Almeida, 2021, p. 32). In addition, the digital realm requires governance mechanisms that promote collaboration among stakeholders and cultivate a democratic and inclusive outlook through institutional arrangements that ensure accountability and transparency. This implies changing the fundamentals of the social contract and introducing mechanisms to facilitate and stimulate the ability to both demand and work towards a clean government (Rothstein, 2021).

Promoting these changes is challenging, especially where structures of power are entrenched in both the online and offline worlds, and corrupt practices flourish in grey areas where certain actions are not entirely illegal or are considered standard business practice. However, continuing to act without aiming for more ambitious goals beyond merely collecting and processing datasets, no matter the limitations they have, and presenting outputs with no guarantee of their use, immunity to ethical risks, or prevention of unexpected negative outcomes is also undesirable. We may continue to have innovative but not lasting and ineffective solutions, although we also still need to determine how to measure the efficiency of ACTs. As we know, corruption is a persistent issue, and technologies quickly become outdated. Unless we rewrite the social contract to ensure that everyone is involved in the governance of the digitalisation of anti-corruption, we risk perpetuating the problems we seek to solve. This holds true not only for ACTs but also for whatever version of the "brave new world" we are already living in.

References

Aarvik, P. (2019). Artificial Intelligence: A promising anti-corruption tool in development settings? *U4 Report*, 1–38. Retrieved from https://www.u4.no/p ublications/artificial-intelligence-a-promising-anti-corruption-tool-in-developm ent-settings.pdf.

Adam, I., and Fazekas, M. (2018). *Are emerging technologies helping win the fight against corruption in developing countries?*Pathways for Prosperity Commission Background Paper Series No. 21.Oxford University Press.

Adam, I., and Fazekas, M. (2021). Are emerging technologies helping win the fight against corruption? A review of the state of evidence. *Information Economics and Policy*, 57, 100950.

Andersen, T.B. (2009). E-Government as an anti-corruption strategy. *Information Economics and Policy*, 21(3), 201–210.

Ang, Y.Y. (2020). *China's Gilded Age: The Paradox of Economic Boom and Vast Corruption*. Cambridge University Press.

Angwin, J., Larson, J., Mattu, S., and Kirchner, L. (2016). Machine bias: There's software used across the country to predict future criminals. and it's biased against blacks. *ProPublica*. Retrieved from https://www.propublica. org/article/machine-bias-risk-assessments-in-criminal-sentencing.

Aranha, A.L. (2020). Lava Jato and Brazil's Web of Accountability Institutions: A Turning Point for Corruption Control? In P. Lagunes and J. Svejnar (Eds.), *Corruption and the Lava Jato Scandal in Latin America*. Routledge.

Avritzer, L., a Filgueiras, F. (2011). *Corrupção e controles democráticos no Brasil*. Ipea.

Bauhr, M., and Grimes, M. (2014). Indignation or resignation: The implications of transparency for societal accountability. *Governance*, 27(2), 291–320.

Benjamin, T., and Couto, F. (2016, December 9). Here's what happened when Brazil banned corporate donations in elections. World Economic Forum. Retrieved from https://www.weforum.org/agenda/2016/12/here-s-what-happ ened-when-brazil-banned-corporate-donations-in-elections/.

Bennett, W.L., and Segerberg, A. (2013) *The Logic of Connective Action: Digital Media and the Personalization of Contentious Politics.* Cambridge University Press.

Berryhill, J., Heang, K.K., Clogher, R., and McBride, K. (2019). Hello, World: Artificial Intelligence and its use in the public sector. OECD Working Papers on Public Governance 36. Organisation for Economic Co-operation and Development. Retrieved from https://www.ospi.es/export/sites/ospi/documents/documentos/Tecnologias-habilitantes/IA-Public-Sector.pdf.

Bertot, J.C., Jaeger, P.T., Munson, S., and Glaisyer, T. (2010). Social Media Technology and Government Transparency. *Computer*, 43(11), 53–59.

Bloom, H. (2004). The story behind the story. In H. Bloom (Ed.), *Bloom's Guides: Aldous Huxley's Brave New World.* Chelsea House Publishers.

BNDES. (2002). Compras governamentais eletrônicas no Brasil: como funcionam os principais sistemas em operação. *Informes BNDES*, 39, April.

Bonini, T., and Trerè, E. (2024). *Algorithms of Resistance: The Everyday Fight against Platform Power.* MIT Press.

Bosi, L., and Uba, K. (2009). The outcomes of social movement. *Mobilization*, 14, 4, 405–411.

Bowen, G.A. (2019). Sensitizing Concepts, In P. Atkinson, S. Delamont, A. Cernat, J.W. Sakshaug, and R.A. Williams (Eds.), *SAGE Research Methods Foundations.* Manchester University Press.

Brelàz, G., Crantschaninov, T.I., and Bellix, L. (2021). Open Government Partnership in São Paulo City and the São Paulo Aberta program: Challenges in the diffusion and institutionalization of a global policy. *Cad. EBAPE.BR*, 19 (1), Jan./March.

Bresser-Pereira, L.C. (1998). Uma reforma gerencial da Administração Pública no Brasil. *Revista do Serviço Público*, 49(1), 5–42.

Brito, J.R. (2009). Breve Histórico do Controle Interno do Poder Executivo Federal: Origem, Evolução, Modelo Atual e Visão de Futuro. *Revista de Negócios*, 7, March. Retrieved from http://judiciary.unifin.com.br/Content/arquivos/20111006173058.pdf.

Bucci, N. (2023, July 7). Robodebt royal commission final report: What did it find and what will happen next? *The Guardian*. Retrieved from https://www.theguardian.com/australia-news/2023/jul/07/robodebt-royal-commission-final-report-what-did-it-find-and-what-will-happen-next.

Bulla, B., and Newell, C. (2020). Sunlight Is the Best Disinfectant: Investigative Journalism in the Age of Lava Jato. In P. Lagunes and J. Svejnar (Eds.), *Corruption and the Lava Jato Scandal in Latin America.* Routledge.

Carson, L., and Mota Prado, M. (2014). *Mapping Corruption & its Institutional Determinants in Brazil.* Retrieved from https://www.brazil4africa.org/wp-content/uploads/publications/working_papers/IRIBA_WP08_Mapping_Corruption_and_its_Institutional_Derminants_in_Brazil.pdf.

Castro Neves, O.M. (2013). *Evolução das políticas de governo aberto no Brasil.* VI Congresso Consad de Gestão Pública. Centro de Convenções Ulysses Guimarães Brasília/DF, April 16, 17 and 18.

Ceva, E., and Jiménez, M.C. (2022). Automating anti-corruption? *Ethics Inf Technol* 24, 48.

Chakraborty, A. (2024). Potentialities and affordances of grassroots civic tech platforms as effective anti-corruption tools: Decoding the story of I Paid a Bribe, India. In A. Mattoni (Ed.), *Digital Media and Grassroots Anti-Corruption.* Edward Elgar.

Charmaz, K. (2006). *Constructing Grounded Theory: A Practical Guide Through Qualitative Analysis.* SAGE.

Charoensukmongkol, P., and Moqbel, M. (2014). Does Investment in ICT Curb or Create More Corruption? A Cross-Country Analysis. *Public Organization Review,* 14(1), 63.

Chemim, R. (2017). *Mãos Limpas e Lava Jato: A corrupção se olha no espelho.* Citadel Grupo Editorial.

Chen, S. (2019, September 22). Is China's corruption-busting AI system "Zero Trust" being turned off for being too efficient? *South China Morning Post.* Retrieved from https://www.scmp.com/news/china/science/article/2184857/chinas-corruption-busting-ai-system-zero-trust-being-turned-being.

Chêne, M. (2012). *Use of Mobile Phones to Detect and Deter Corruption.* Department for International Development. Retrieved from https://www.gov.uk/dfid-research-outputs/use-of-mobile-phones-to-detect-and-deter-corruption.

Clarke, A., Friese, C., and Washburn, R. (Eds.). (2015). *Situational Analysis in Practice: Mapping Research with Grounded Theory.* Left Coast Press.

Couldry, N. (2012). *Media, Society, World: Social Theory and Digital Media Practice.* Polity Press.

Coutinho, G.L. (2012). Aniita: Uma abordagem pragmática para o gerenciamento de risco aduaneiro baseada em software. *Prêmio de Criatividade e Inovação da RFB.* Retrieved from https://repositorio.enap.gov.br/bitstream/1/4607/1/Mencao%20honrosa%20do%2011%C2%BA%20Premio%20RFB.pdf (accessed on April 23, 2021).

CPMI. (2006). Relatório dos Trabalhos da CPMI das Ambulâncias. Senado do Brasil. Retrieved from https://www.senado.leg.br/comissoes/CPI/Ambulancias/CPMI_RelatorioParcial_1.pdf.

Cristóvam, J.S.S., Saikali, L.B., and Sousa, T.P. (2020) Governo Digital na Implementação de Serviços Públicos para a Concretização de Direitos Sociais no Brasil. *Seqüência (Florianópolis),* 84, 209–242.

Da Ros, L., and Taylor, M.M. (2022). *Brazilian Politics on Trial: Corruption and Reform under Democracy.* Lynne Rienner.

Davies, T., and Fumega, S. (2014). Mixed incentives: Adopting ICT innovations for transparency, accountability, and anti-corruption. *U4 Issue (4).* Retrieved from http://www.cmi.no/publications/file/5172-mixedincentives.pdf.

Department of Justice. (2016). Odebrecht and Braskem Plead Guilty and Agree to Pay at Least $3.5 Billion in Global Penalties to Resolve Largest Foreign Bribery Case in History. US Department of Justice Office of Public Affairs. Retrieved from https://www.justice.gov/opa/pr/odebrecht-and-bra skem-plead-guilty-and-agree-pay-least-35-billion-global-penalties-resolve.

Earl, J., and Kimport, K. (2011). Where We Have Been and Where We Are Headed. In *Digitally Enabled Social Change: Activism in the Internet Age.* MIT Press.

Edler-Duarte, D. (2021). The Making of Crime Predictions: Sociotechnical Assemblages and the Controversies of Governing Future Crime. *Surveillance & Society,* 19(2): 199–215.

Elbahnasawy, N.G. (2014). E-Government, Internet Adoption, and Corruption: An Empirical Investigation. *World Development,* 57, 114–126.

Elliott-Negri, L., Jabola-Carolus, I., Jasper, J., Mahlbacher, J., Weisskircher, M., and Zhelnina, A. (2021). Social Movement Gains and Losses: Dilemmas of Arena Creation. *Partecipazione e Conflitto,* 14(3).

Fernández-Vázquez, P., Barberá, P., and Rivero, G. (2015). Rooting out corruption or rooting for corruption? The heterogeneous electoral consequences of scandals. *Political Science Research and Methods,* 4(2), 379–397.

Figueiredo, A.C. (2010). The Collor Impeachment and Presidential Government in Brazil. In M. Llanos and L. Marsteintredet (Eds.), *Presidential Breakdowns in Latin America: Causes and Outcomes of Executive Instability in Developing Democracies.* Palgrave Macmillan.

Filgueiras, F. (2023). Designing artificial intelligence policy: Comparing design spaces in Latin America. *Latin American Policy,* 14(1), 1–17.

Filgueiras, F., and Almeida, V. (2021). *Governance for the Digital World: Neither More State nor More Market.* Palgrave Macmillan.

Folha de S.Paulo. (2000, September 30). Entenda o caso da feira de Hannover. Retrieved from https://www1.folha.uol.com.br/folha/brasil/ult96u7405.shtml.

Folha de S.Paulo. (2003, July 27). Privatização das teles foi seguida por escândalos. Retrieved from https://www1.folha.uol.com.br/folha/dinheiro/ult91u70988.shtml.

Folha de S.Paulo. (2009, August 20). Câmara terá de entregar notas fiscais sobre gastos da verba indenizatória à Folha. Retrieved from https://www1.folha.uol.com.br/folha/brasil/ult96u612372.shtml.

France, G. (2019). Brazil: Setbacks in the legal and institutional anti-corruption frameworks. Transparency International. Retrieved from https://www.transparency.org/en/publications/brazil-setbacks-in-the-legal-and-institutional-anti-corruption-frameworks.

Freire, M. (2015, April 1). *Conheça dez histórias de corrupção durante a ditadura militar.* UOL. Retrieved from https://noticias.uol.com.br/politica/ultimas-noticias/2015/04/01/conheca-dez-historias-de-corrupcao-durante-a-ditadura-militar.htm.

Freire, D., Galdino, M., and Mignozzetti, U. (2020). Bottom-up accountability and public service provision: Evidence from a field experiment in Brazil. *Research and Politics*, 7(2), 1–8.

Freitas, J. (1987, May 13). Concorrência da ferrovia Norte-Sul foi uma farsa. *Folha de S.Paulo*. Retrieved from https://www1.folha.uol.com.br/folha/80a nos/marcos_do_jornalismo-03.shtml.

Friend, C. (2024). Social Contract Theory. *Encyclopedia of Philosophy* (online). Retrieved from https://iep.utm.edu/soc-cont/ (accessed on March 4, 2024).

Gaetani, F. (2005). Estratégia e Gestão da Mudança nas Políticas de Gestão Pública. In E. Levy and P.A. Drago (Eds.), *Gestão pública no Brasil contemporâneo*. FUNDAP.

Gainty, C. (2023, January 16). From a "deranged" provocateur to IBM's failed AI superproject: The controversial story of how data has transformed healthcare. *The Conversation*.

Galdino, M., Mondo, B.V., Sakai, J.M., and Paiva, N. (2023, November 3). The Civil Society Organizations effect: A mixed-methods analysis of bottom-up accountability in Brazilian public policy. SocArXiv. Retrieved from https://doi.org/10.31235/osf.io/s82dn.

Gaspar, M. (2020). *A Organização: A Odebrecht e o esquema de corrupção que chocou o mundo*. Companhia das Letras.

Glaser, B.G., and Strauss, A.L. (1967). *The Discovery of Grounded Theory*. Aldine.

Gohn, M.G. (2014). A sociedade brasileira em movimento: Vozes das ruas e seus ecos políticos e sociais. *Caderno CRH*, 27(71), 431–441.

Gomes, C.F.S., and Costa, H.G. (2015). Aplicação de métodos multicritério ao problema de escolha de modelos de pagamento eletrônico por cartão de crédito. *Production*, 25(1), 54–68.

Governo do Brasil (2023, June 16). Gestão apresenta propostas de expansão da plataforma GOV.BR a secretários estaduais de administração. Retrieved from https://www.gov.br/gestao/pt-br/assuntos/noticias/2023/junho/gestao-ap resenta-propostas-de-expansao-da-plataforma-gov-br-a-secretarios-estadua is-de-administracao.

Graft, A., Verhulst, S., and Young, A. (2016). Brazil's open budget transparency portal: Making public how public money is spent. *Report Gov Lab – Open Data's Impact*. Retrieved from https://odimpact.org/files/ca se-study-brazil.pdf.

Grimes, M. (2008). *The conditions of successful civil society involvement in combating corruption: A survey of case study evidence*. QoG Working Paper Series, 22.

Gurin, J. (2014). Open Governments, Open Data: A New Lever for Transparency, Citizen Engagement, and Economic Growth. *The SAIS Review of International Affairs*, 34(1), 71–82.

Higgins, E. (2021). *We Are Bellingcat: An Intelligence Agency for the People*. Bloomsbury.

Holdo, M. (2019). Cooptation and non-cooptation: Elite strategies in response to social protest. *Social Movement Studies*, 18(4), 444–462.

Huss, O. (2020). *How Corruption and Anti-Corruption Policies Sustain Hybrid Regimes: Strategies of Political Domination Under Ukraine's Presidents in 1994–2014.* Columbia University Press.

Huxley, A. (1932). *Brave New World.* Penguin.

Iqbal, M.S., and Seo, J.W. (2008). E-governance as an anti-corruption tool: Korean cases. *Journal of the Korean Association for Regional Information Society,* 11(2), 51–78.

Jambreiro Filho, J. (2015a). Artificial Intelligence in the Customs Selection System through Machine Learning (Sisam). Retrieved from https://www.jambeiro.com.br/jorgefilho/sisam_mono_eng.pdf (accessed on September 29, 2022).

Jambreiro Filho, J. (2015b). Inteligência Artificial no Sistema de Seleção Aduaneira por Aprendizado de Máquina. In *Prêmio de Criatividade e Inovação da RFB,* 14th edn. Receita Federal do Brasil, Escola de Administração Fazendária.

Jambreiro Filho, J. (2019). Artificial Intelligence Initiatives in the Special Secretariat of Federal Revenue of Brazil. Retrieved from https://www.jambeiro.com.br/jorgefilho/AI_Brazil_Federal%20Revenue%20_2019.pdf (accessed on September 29, 2022).

Jefferson, B.J. (2018). Predictable Policing: Predictive Crime Mapping and Geographies of Policing and Race. *Annals of the American Association of Geographers,* 108(1), 1–16.

Jha, C.K., and Sarangi, S. (2017). Does social media reduce corruption? *Information Economics and Policy,* 39, 60–71.

Johnston, M. (2005). *Syndromes of Corruption: Wealth, Power, and Democracy.* Cambridge University Press.

JOTA. (2019, September 10). *JOTA lança aprovômetro de projetos legislativos.* Retrieved from https://www.jota.info/dados/jota-lanca-aprovometro-de-projetos-legislativos-10092019.

Kim, K., and Kang, T. (2019). Will Blockchain Bring an End to Corruption? *International Journal of Information Systems and Social Change,* 10 (2), 35–44.

Knaus, C., and Henriques-Gomes, L. (2023, May 3). How two reporters exposed Centrelink's robodebt injustice and gave voice to the victimised. *The Guardian.* Retrieved from https://www.theguardian.com/media/2023/may/04/how-two-reporters-exposed-centrelinks-robodebt-injustice-and-gave-voice-to-the-victimised.

Köbis, N., Starke, C., and Edward-Gill, J. (2022a). The Corruption Risks of Artificial Intelligence. Working Paper.Transparency International. Retrieved from https://knowledgehub.transparency.org/assets/uploads/kproducts/The-Corruption-Risks-of-Artificial-Intelligence.pdf.

Köbis, N., Starke, C., and Rahwan, I. (2022b). The promise and perils of using artificial intelligence to fight corruption. *Nat Mach Intell,* 4, 418–424.

Kossow, N. (2020a). Digital Anti-Corruption: Hopes and Challenges. In A. Mungiu-Pippidi and P. Heywood (Eds.), *A Research Agenda for Studies of Corruption*. Edward Elgar.

Kossow, N. (2020b). *Digitizing Collective Action: How Digital Technologies Support Civil Society's Struggle against Corruption*. PhD Dissertation, Hertie School, Germany.

Kossow, N., and Dykes, V. (2018). Bitcoin, Blockchain and Corruption: An Overview. Transparency International. Retrieved from https://knowledgehub. transparency.org/helpdesk/bitcoin-blockchain-andcorruption-an-overview. 106

Kossow, N., and Kukutschka, R.M.B. (2017). Civil society and online connectivity: controlling corruption on the net? *Crime, Law and Social Change*, 68 (4), 459–476.

Kukutschka, R.M.B. (2016). Technology against corruption: the potential of online corruption- reporting apps and other platforms. *U4 Anti-Corruption Centre*. Retrieved from https://www.u4.no/publications/technology-against-corruption-the-potential-of-online-corruption-reporting-apps-and-other-platforms.

Lagunes, P., Michener, G., Odilla, F., and Pires, B. (2021a). President Bolsonaro's Promises and Actions on Corruption Control. *Revista Direito GV*, 17(2).

Lagunes, P., Odilla, F., and Svejnar, J. (2021b). *Corrupção e o escândalo da Lava Jato na América Latina*. Editora FGV.

Lascoumes, P. and Le Galès, P. (2007). Introduction: Understanding Public Policy through Its Instruments – From the Nature of Instruments to the Sociology of Public Policy Instrumentation. *Governance*, 20(1), 1–21.

Latinobarometro. (2020). *Online analysis*. Retrieved from https://www.latinobarometro.org/latOnline.jsp.

Lima, O.D.W., and Andrade, N. (2019). Fairness in Risk Estimation of Brazilian Public Contracts. *Symposium on Knowledge Discovery, Mining and Learning, KDMILE 2019 – Applications Track*. Retrieved from https://sol.sbc.org.br/index.php/kdmile/article/view/8789/8690.

Long, W. (1988, April 26). Peril for Democracy: Brazil Reels Under Tales of Corruption. *Los Angeles Times*. Retrieved from https://www.latimes.com/archives/la-xpm-1988-04-26-mn-1595-story.html.

Lopes, A.A.L. (2018). A Evolução do SIAFI Enquanto Sistema de Controle Interno do Governo Federal. *Revista Científica Multidisciplinar Núcleo do Conhecimento, Ano* 03, 7(4), 40–50.

Mainwaring, S., and Welna, C. (Eds). (2003) *Democratic Accountability in Latin America*. Oxford University Press.

Matais, A.*et al.* (2016, December 12). Anões do Orçamento fizeram Odebrecht mudar estratégia no Congresso, diz delator. *O Estado de S.Paulo*. Retrieved from https://noticias.uol.com.br/ultimas-noticias/agencia-estado/2016/12/12/anoes-do-orcamento-fizeram-odebrecht-mudar-estrategia-no-congresso-diz-delator.htm.

Mattoni, A. (2017). From data extraction to data leaking: Data-activism in Italian and Spanish anti-corruption campaigns. *Partecipazione e Conflitto*, 10(3), 723–746.

Mattoni, A. (2020). The grounded theory method to study data-enabled activism against corruption: Between global communicative infrastructures and local activists' experiences of big data. *European Journal of Communication*, 35(3), 265–277.

Mattoni, A. (2021). Digital Media in Grassroots Anti-Corruption Mobilizations. In D.A. Rohlinger and S. Sobieraj (Eds.), *The Oxford Handbook of Sociology and Digital Media*. Oxford University Press.

Mattoni, A. (2024). Digital Media and Anti-Corruption Technologies from the Grassroots: an Introduction. In A. Mattoni (Ed.), *Digital Media and Grassroots Anti-corruption*. Edward Elgar.

Mattoni, A., and Odilla, F. (2021). Digital media, activism, and social movements' outcomes in the policy arena. The case of two anti-corruption mobilizations in Brazil. *Partecipazione e Conflitto*, 7623(14), 1127–1150.

McCormack, C. (2016). *Democracy rebooted: The future of technology in elections*. Atlantic Council. Retrieved from https://www.atlanticcouncil.org/in-depth-research-reports/report/democracy-rebooted-the-future-of-technology-in-elections-report/.

Mendes, L. (2022, September 17). Lei da Ficha Limpa barra ao menos 185 candidaturas. *Poder 360*. Retrieved from https://www.poder360.com.br/eleicoes/lei-da-ficha-limpa-barra-ao-menos-185-candidaturas/.

Meneghetti, D. (2017, October 20). A origem de 35 expressões populares brasileiras. *Super Interessante*. Retrieved from https://super.abril.com.br/especiais/nao-marque-touca-a-origem-de-35-expressoes-populares/https://super.abril.com.br/especiais/nao-marque-touca-a-origem-de-35-expressoes-populares/.

Ministério da Transparência e Controladoria-Geral da União. (2018) Transparency Against Corruption: The Brazilian Experience. Retrieved from ttps://repositorio.cgu.gov.br/handle/1/27534.

Monaco, N., and Woolley, S. (2022). *Bots*. Polity Press.

Montevechi, C. (2021). Ativismo Anticorrupção no Brasil e a Teoria dos Movimentos Sociais. *Revista Brasileira de Ciência Política*, 34(235262), 1–37.

Morris, S. D. (2021). *The Corruption Debates: Left vs. Right – and Does It Matter – in the Americas*. Lynne Rienner Publishers.

Mota Prado, M., and Cornelius, E. (2020). Institutional Multiplicity and the Fight Against Corruption: A Research Agenda for the Brazilian Accountability Network. *Rev. direito GV*, 16(3).

Mota Prado, M., Carson, L.D., and Correa, I. (2015). The Brazilian Clean Company Act: Using Institutional Multiplicity for Effective Punishment. *Osgoode Legal Studies Research Paper No. 48/2015*. Retrieved from https://papers.ssrn.com/sol3/papers.cfm?abstract_id=2673799.

Mungiu-Pippidi, A., and Fazekas, M. (2020). How to define and measure corruption. In A. Mungiu-Pippidi and P. Heywood (Eds.), *A Research Agenda for Studies of Corruption*. Edward Elgar.

Murray, A., David-Barrett, E., and Ceballos, J.C. (2023). *Country Insights Brief Brazil.* Insights Series 03, February. IACA. Retrieved from www.iaca.int/mea suring-corruption/wp-content/uploads/2023/02/GPMC_Brazil_Insights_Brief_ 20022023_online_1.pdf.

Neves, F., and Silva, P.B. (2023). From Paper to Digital: e-Governments' Evolution and Pitfalls in Brazil. In C. Gaie and M. Mehta (Eds.), *Recent Advances in Data and Algorithms for e- Government*, vol. 5. AISSE, p. 193.

Neves, F., Silva, P.B., and Carvalho, H.L.M. (2019). Artificial Ladies against Corruption: Searching for Legitimacy at the Brazilian Supreme Audit Institution. *Revista de Contabilidade e Organizações*, 13, 31–50.

Nishijima, M., Ivanauskas, T.M., and Sarti, F.M. (2017). Evolution and determinants of digital divide in Brazil (2005–2013). *Telecommunications Policy*, 41(1), 12–24.

Noveck, B. S., Koga, K., Garcia, R. A., Deleanu, H., and Cantú-Pedraza, D. (2018). *Smarter Crowdsourcing for Anti-corruption: A Handbook of Innovative Legal, Technical, and Policy Proposals and a Guide to their Implementation.* Inter-American Development Bank. Retrieved from http s://publications.iadb.org/en/smarter-crowdsourcinganti-corruption-handboo k-innovative-legal-technical-and-policy-proposalsand.

O'Donnell, G. (1998). Horizontal accountability and new polyarchies. *Lua Nova: Revista de Cultura e Política*, 44, 27–54.

O'Donnell, G. (1999). Horizontal accountability in new democracies. In A. Schedler, L. Diamond, and M. Plattner (Eds.), *The Self-Restraining State: Power and Accountability in New Democracies.* Lynne Rienner.

Odilla, F. (2020a). Oversee and Punish: Understanding the Fight Against Corruption Involving Government Workers in Brazil. *Politics and Governance*, 8(2), 1–13.

Odilla, F. (2020b). *Oversee & Punish: Understanding the fight against corruption involving government workers in the federal executive branch in Brazil.* PhD Thesis. King's College London.

Odilla, F. (2023a). Bots against corruption: Exploring the benefits and limitations of AI-based anti-corruption technology. *Crime Law Soc Change* 80, 353–396.

Odilla, F. (2023b). *Unfairness in AI anti-corruption tools: Main drivers and consequences.* Working Paper. Presented at the 2023 ECPR General Conference, September 4–8, Prague.

Odilla, F. (2024). From concerned citizens to civic bots: The bottom-up fight against corruption in Brazil from a longitudinal perspective. In A. Mattoni (Ed.) *Digital Media and Grassroots Anti-Corruption.* Edward Elgar.

Odilla, F., and Rodriguez-Olivari, D. (2021). Corruption control under fire: A brief history of Brazil's Office of the Comptroller General. In J. Pozsgai-Alvarez (Ed.) *The Politics of Anti-Corruption Agencies in Latin America.* Routledge.

Odilla, F., and Mattoni, A. (2023). Unveiling the layers of data activism: The organising of civic innovation to fight corruption in Brazil. *Big Data & Society*, 10(2).

Odilla, F., and Veloso, C. (2024). Citizens and their bots that sniff corruption: Using digital media to expose politicians who misuse public money. *American Behavioral Scientist*.

Oliveira, L.G.L. (2017). Dez anos de CNJ: Reflexões do envolvimento com a melhoria da eficiência do judiciário brasileiro. *Revista do Serviço Público*, 68 (3), 631–656.

Organisation for Economic Co-operation and Development (OECD). (2016). *The Korean Public Procurement Service Innovating for Effectiveness*. Retrieved from https://www.oecdilibrary.org/governance/the-korean-public-procurementservice_9789264249431-en.

Organisation for Economic Co-operation and Development (OECD). (2020). *Going digital in Brazil*. Retrieved from https://www.oecd-ilibrary.org/sites/c5840db0-en/index.html?itemId=/content/component/c5840db0-en#biblio-d1e22611.

Otranto Alves, L.C., Soares Silva, A., and da Fonseca, A.C.P.D. (2008). Implicações da Adoção do Modelo de Merchant na Avaliação do Uso da TI para Controle Gerencial do Serviço Público: Análise do Portal Comprasnet. *Contab. Vista & Rev.*, 19(1), 83–108.

Pedrozo, S. (2013). New Media Use in Brazil: Digital Inclusion or Digital Divide? *Online Journal of Communication and Media Technologies*, 3(1).

Pereyra, S. (2019). Corruption Scandals and Anti-Corruption Policies in Argentina. *Journal of Politics in Latin America*, 11(3), 348–361.

Petherick, A. (2015, August 14). Brazil and the Bloodsuckers. *Foreign Policy*. Retrieved from https://foreignpolicy.com/2015/08/14/brazil-and-the-bloodsuckers-corruption-lottery/.

Picci, L. (2024). *Rethinking Corruption*. Cambridge University Press.

Pires, B. (2022, July). Farra ilimitada. *Revista Piauí. Edição* 190. https://piaui.folha.uol.com.br/materia/farra-ilimitada/.

Piven, F.F., and Cloward, R. (1979). *Poor People's Movements: Why They Succeed, How They Fail*. Knopf Doubleday Publishing Group.

Pogrebinschi, T. (2018). Experimenting with Participation and Deliberation: Is Democracy Turning Pragmatic? In T. Falleti and E. Parrado (Eds.), *Latin America Since the Left Turn*. University of Pennsylvania Press.

Portela, M. (2022, February 7). CGU identifica fraude de R$ 809,9 milhões no auxílio emergencial. *Correio Braziliense*. Retrieved from https://www.correiobraziliense.com.br/economia/2022/02/4983309-cgu-identifica-fraude-de-rs-8099-milhoes-no-auxilio-emergencial.html.

Power, T.J., and Taylor, M.M. (2011). Introduction: Accountability Institutions and Political Corruption in Brazil. In T.J. Power and M.M. Taylor (Eds.), *Corruption and Democracy in Brazil: The Struggle for Accountability*. University of Notre Dame.

Praça, S. (2011). Corrupção e reforma institucional no Brasil, 1988–2008. *Opinião Pública*, 17(1), 137–162.

Praça, S., and Taylor, M.M. (2014). Inching Toward Accountability: The Evolution of Brazil's Anticorruption Institutions, 1985–2010. *Latin American Politics and Society*, 56(2), 27–48.

Rezende, C. (2021, June 29). Governo Bolsonaro pediu propina de US$ 1 por dose, diz vendedor de vacina. *Folha de S.Paulo*. Retrieved from https://www1.folha.uol.com.br/poder/2021/06/exclusivo-governo-bolsonaro-pediu-p ropina-de-us-1-por-dose-diz-vendedor-de-vacina.shtml.

Rich, J. (2019). *State-Sponsored Activism: Bureaucrats and Social Movements in Democratic Brazil*. Cambridge University Press.

Rich, J. (2020). Organizing Twenty-First-Century Activism: From Structure to Strategy in Latin American Social Movements. *Latin American Research Review*, 55(3), 430–444.

Rodrigues, F. (2014,June 16). *Conheça a história da compra de votos a favor da emenda da reeleição*. Blog do Fernando Rodrigues, Brasília. Retrieved from https://fernandorodrigues.blogosfera.uol.com.br/2014/06/16/conheca -a-historia-da-compra-de-votos-a-favor-da-emenda-da-reeleicao/ (accessed on April 10, 2020).

Rodrigues, F., and Lobato, E. (1999, May 25). FHC tomou partido de consórcio no leilão das teles, revelam fitas. *Folha de S.Paulo*. Retrieved from https://www1.folha.uol.com.br/fsp/brasil/especial/sp7.htm (accessed on April 19, 2020).

Rose-Ackerman, S. (1996). Democracy and "Grand" Corruption. In R. Williams (Ed.), *Explaining Corruption: The Politics of Corruption*. Edward Elgar.

Rose-Ackerman, S. (2021). Corruption and Covid-19. *Eunomía. Revista en Cultura de la Legalidad*, 20, 16–13.

Rothstein, B. (2021). The Social Contract and the Indirect Approach to Anti-Corruption. In *Controlling Corruption: The Social Contract Approach*. Oxford University Press.

Sadek, M.T.A. (2019). Combate à Corrupção: novos tempos. *Revista da CGU*, 11(20), 1276–1283.

Saldanha, P. (2022, April 6). Governo Bolsonaro destina R$ 26 mi em kit robótica para escolas sem água e computador. *Folha de S. Paulo*. Retrieved from https://www1.folha.uol.com.br/poder/2022/04/governo-bolsonaro-des tina-r-26-mi-em-kit-robotica-para-escolas-sem-agua-e-computador.shtml.

Sano, H. (2020). Laboratórios de Inovação no Setor Público: mapeamento e diagnóstico de experiências nacionais. *Cadernos Enap 69*.

Saraiva, A. (2018). *A implementação do SEI – Sistema Eletrônico de Informações*. Enap Casoteca de Gestão Pública.

Savaget, P., Chiarini, T., and Evans, S. (2019). Empowering political participation through artificial intelligence. *Science and Public Policy*, 46(3), 369–380.

Schwarcz, L. (2019). *Sobre o Autoritarismo Brasileiro*. Companhia das Letras.

Senne, F. (2021). Beyond connectivity: Internet for all. *Internet Sectoral Overview*, 2(June).

Siegel, E. (2018,February 19). How to Fight Bias with Predictive Policing . *Scientific American* (blog). Retrieved from https://blogs.scientificamerican. com/voices/how-to-fight-bias-with-predictive-policing/.

Smulovitz, C., and Peruzzotti, E. (2000). Societal Accountability in Latin America. *The Journal of Democracy*, 11(4), 147–158.

Starke, C., Kieslich, K., Reichert, M., and Köbis, N. (2023,January 13). Algorithms against Corruption: A Conjoint Study on Designing Automated Twitter Posts to Encourage Collective Action. SocArXiv.

Sturges, P. (2004). Corruption, transparency and a role for ICT? *International Journal of Information Ethics*, 2(11).

Taylor, M.M., and Buranelli, V.C. (2007). Ending up in Pizza: Accountability as a Problem of Institutional Arrangement in Brazil. *Latin American Politics and Society*, 49(1), 59–87.

Transparency International. (2012). 15th International Anti-Corruption Conference closes with an urgent call to end impunity. Retrieved from https://www.transparency.org/en/press/20121110-15th-international-anti-corruption-conference-closes-with-an-urgen.

Tribunal de Contas da União (TCU). (2014). TCU e a Informatização. In *Brasil, Tribunal de Contas da União. Tribunal de Contas da União: Evolução histórica e administrativa.* TCU. Retrieved from https://portal.tcu.gov.br/cen tro-cultural-tcu/museu-do-tribunal-de-contas-da-uniao/tcu-a-evolucao-do-c ontrole/tcu-e-a-informatizacao.htm (accessed on August 18, 2023).

Tribunal de Contas da União (TCU). (2016). *Fiscobras: 20 anos. Tribunal de Contas da União.* TCU, Secretaria-Geral de Controle Externo. Retrieved from https://portal.tcu.gov.br/data/files/93/C4/3D/41/F6DEF610F5680BF6F18818A 8/Fiscobras_20_anos.pdf .

Tribunal de Contas da União (TCU). (2022). *Acórdão 1139/2022.* Retrieved from https://pesquisa.apps.tcu.gov.br/#/documento/acordao-completo/6662 20218.PROC/%2520/DTRELEVANCIA%2520desc%252C%2520NUMA CORDAOINT%2520desc/0/%2520.

TSE. (2014). Conheça a história da urna eletrônica brasileira, que completa 18 anos. Retrieved from https://www.tse.jus.br/comunicacao/noticias/2014/Junho/ conheca-a-historia-da-urna-eletronica-brasileira-que-completa-18-anos.

United Nations Development Programme (UNDP). (2011). *Study on the Role of Social Media for Enhancing Public Transparency and Accountability in Eastern Europe and the Commonwealth of Independent States: Emerging Models, Opportunities and Challenges.* UNDP Regional Office for Eastern Europe & the CIS. Retrieved from https://issuu.com/undp_in_europe_cis/ docs/social_media_report_-_external.

Urquhart, C. (2013). *Grounded Theory for Qualitative Research.* SAGE.

van Bekkum, M., and Borgesius, F.Z. (2021). Digital welfare fraud detection and the Dutch SyRI judgment. *European Journal of Social Security*, 23(4), 323–340.

van Dijck, J., Poell, T., and De Wall, M. (2018). *The Platform Society: Public Values in a Connective World.* Oxford University Press.

Varraich, A. (2014). *Corruption: An Umbrella Concept.* Retrieved from https:// qog.pol.gu.se/digitalAssets/1551/1551604_2014_05_varraich.pdf.

Vieira, J., and Miranda, L.F. (forthcoming). When corruption strikes back: how is congress reversing anti-corruption reforms in Brazil. In F. Odilla and K. Tsimonis (Eds.) *Corruption and Anti-Corruption Upside Down: New Perspectives from the Global South.* Palgrave Macmillan.

Wickberg, S. (2018). The role of mediated scandals in the definition of anti-corruption norms. In I. Kubbe and A. Engelbert (Eds.), *Corruption and Norms*. Springer.

Zinnbauer, D. (2015). Crowdsourced Corruption Reporting: What Petrified Forests, Street Music, Bath Towels, and the Taxman Can Tell Us About the Prospects for Its Future. *Policy & Internet*, 7(1), 1–24.

Index

For Product Safety Concerns and Information please contact our EU
representative GPSR@taylorandfrancis.com
Taylor & Francis Verlag GmbH, Kaufingerstraße 24, 80331 München, Germany